T0201768

ARCHITECTING THE TELECOMMUNICATION EVOLUTION

OTHER TELECOMMUNICATIONS BOOKS FROM AUERBACH

Active and Programmable Networks for Adaptive Architectures and Services
Syed Asad Hussain
ISBN: 0-8493-8214-9

Business Strategies for the Next-Generation Network
Nigel Seel
ISBN: 0-8493-8035-9

Chaos Applications in Telecommunications
Peter Stavroulakis
ISBN: 0-8493-3832-8

Context-Aware Pervasive Systems: Architectures for a New Breed of Applications
Seng Loke
ISBN: 0-8493-7255-0

Fundamentals of DSL Technology
Philip Golden, Herve Dedieu, Krista S Jacobsen
ISBN: 0-8493-1913-7

IP Multimedia Subsystem: Service Infrastructure to Converge NGN, 3G and the Internet
Rebecca Copeland
ISBN: 0-8493-9250-0

Mobile Middleware
Paolo Bellavista and Antonio Corradi
ISBN: 0-8493-3833-6

MPLS for Metropolitan Area Networks
Nam-Kee Tan
ISBN: 0-8493-2212-X

Network Security Technologies, Second Edition
Kwok T Fung
ISBN: 0-8493-3027-0

Performance Modeling and Analysis of Bluetooth Networks: Polling, Scheduling, and Traffic Control
Jelena Misic and Vojislav B Misic
ISBN: 0-8493-3157-9

Performance Optimization of Digital Communications Systems
Vladimir Mitlin
ISBN: 0-8493-6896-0

A Practical Guide to Content Delivery Networks
Gilbert Held
ISBN: 0-8493-3649-X

Resource, Mobility and Security Management in Wireless Networks and Mobile Communications
Yan Zhang, Honglin Hu, and Masayuki Fujise
ISBN: 0-8493-8036-7

Service-Oriented Architectures in Telecommunications
Vijay K Gurbani, Xian-He Sun
ISBN: 0-8493-9567-4

TCP Performance over UMTS-HSDPA Systems
Mohamad Assaad and Djamal Zeghlache
ISBN: 0-8493-6838-3

Testing Integrated QoS of VoIP: Packets to Perceptual Voice Quality
Vlatko Lipovac
ISBN: 0-8493-3521-3

Traffic Management in IP-Based Communications
Trinh Anh Tuan
ISBN: 0-8493-9577-1

Understanding Broadband over Power Line
Gilbert Held
ISBN: 0-8493-9846-0

WiMAX: A Wireless Technology Revolution
G.S.V. Radha Krishna Rao, G. Radhamani
ISBN: 0-8493-7059-0

WiMAX: Taking Wireless to the MAX
Deepak Pareek
ISBN: 0-8493-7186-4

Wireless Mesh Networks
Gilbert Held
ISBN: 0-8493-2960-4

Wireless Security Handbook
Aaron E Earle
ISBN: 0-8493-3378-4

AUERBACH PUBLICATIONS
www.auerbach-publications.com
To Order Call: 1-800-272-7737 • Fax: 1-800-374-3401
E-mail: orders@crcpress.com

ARCHITECTING THE TELECOMMUNICATION EVOLUTION

Toward Converged Network Services

Vijay K. Gurbani
Xian-He Sun

Auerbach Publications
Taylor & Francis Group
Boca Raton   New York

Auerbach Publications is an imprint of the
Taylor & Francis Group, an informa business

Auerbach Publications
Taylor & Francis Group
6000 Broken Sound Parkway NW, Suite 300
Boca Raton, FL 33487-2742

International Standard Book Number-10: 0-8493-9567-4 (Hardcover)
International Standard Book Number-13: 978-0-8493-9567-3 (Hardcover)

Library of Congress Cataloging-in-Publication Data

Gurbani, Vijay K.
    Architecting the telecommunication evolution : toward converged network services / Vijay K. Gurbani and Xian-He Sun.
      p. cm.
    ISBN 0-8493-9567-4 (alk. paper)
    1. Telecommunication systems. 2. Computer networks. I. Sun, Xian-He. II. Title.

TK5101.G79 2006
384--dc22   2006045976

Visit the Taylor & Francis Web site at
http://www.taylorandfrancis.com

and the Auerbach Web site at
http://www.auerbach-publications.com

Dedication

To my parents, for their love; to my wife, for her understanding;
to my children, for their future.

Vijay K. Gurbani

To my parents, Yu Lin (林雨) and Chang-Xiang Sun (孙昌湘), my wife,
Hong Zhang (张红), and my children, Linda and Alan.

Xian-He Sun, (孙贤和**)**

Table of Contents

Foreword

Internet Telephony can be defined as the conversational exchange of voice or more generally multimedia data over packet switched networks. It is no longer novel. The first two-party voice call over a packet switched network was made in the '70s as part of the Network Secure Communications project run by the Advanced Research Projects Agency (ARPA). After this first experiment, a handful of proprietary systems were developed in the '80s. It is only in the late '90s and early '00s that wide spread standards came to life.

Two main sets of standards have emerged: H.323 from the ITU-T and the session initiation protocol (SIP) from the IETF. SIP is certainly now the prime set of standards because it has been adopted by the main telecommunications standards bodies including the Third Generation Partnership Projects (3GPP/3GPP2). 3GGP and 3GPP2 are the bodies that are specifying the third generation mobile networks. Although SIP is today the most prominent set of Internet Telephony standards, H.323 remains a prominent standard due to its installed base, especially in the enterprise environments.

Two-party voice call is the basic service provided by Internet Telephony service providers. Value added services, or more simply services, can be defined as anything that go beyond this basic service. These are critical to the success and survival of telephony service providers. Different paradigms are used in the public switched telephony network (PSTN) and in Internet Telephony for providing these services. On the one hand, the intelligent network (IN) is the main paradigm used in the PSTN. On the other hand, SIP servlets, and SIP CGI are examples of paradigms used in Internet Telephony.

Evolving from the current PSTN to Internet Telephony is not an easy task. A possible starting point is to enrich the set of PSTN services by taking advantage of the most widely deployed packet switched infrastructure today: the Internet. Services such as the well-known Internet call waiting can be deployed in this manner. The IETF has realized this and has produced a protocol in that regard, the SPIRITS (Services in PSTN requesting Internet Services) protocol. In much the same way services originating in the PSTN use services on the Internet, services originating on the Internet can also use services in the PSTN. A well-known example, click to dial, shows another way of enriching PSTN services by taking advantage of the Internet.

This book focuses on how the evolution can be architected. It introduces Internet Telephony, provides background information on how value added services can be engineered in both traditional circuit switched telephony and Internet Telephony. It then dives into how the set of PSTN services can be significantly enhanced by taking advantage of the Internet. The services originating in the Internet and using PSTN services and services originating in the PSTN and using the Internet are successively presented.

The two authors are uniquely qualified to explain how this evolution can be architected. Vijay K. Gurbani is a co-author of the two RFCs produced by the IETF on the topic. In addition, he has published several papers on the same topic in prestigious journals and conference proceedings. Xian-He Sun is also a preeminent researcher in the field of high performance computing and communication. Both authors have used their unique expertise to produce this very first book on the topic.

Roch Glitho, Ph.D.
Editor-in-Chief, *IEEE Communications Magazine,* 2003–2005
Technical Expert, Service Layer Technologies/Ericsson Canada
Concordia University, Montreal, Canada

The Authors

Vijay K. Gurbani is a distinguished member of the technical staff in the Security Technology Research Group in Bell Laboratories, the research division of Lucent Technologies, Inc. He holds a Ph.D. in Computer Science from the Illinois Institute of Technology, Chicago, Illinois, and a M.Sc. and B.Sc., both in Computer Science, from Bradley University, Peoria, Illinois. Vijay's current work focuses on security aspects of next generation signaling protocols. His earlier work involved the use of SIP-based services which use the public switched telephone network (PSTN) and the Internet. He is the author of two Internet Engineering Task Force RFCs which use SIP as a facilitator for services spanning the Internet and the PSTN. His research interests are Internet telephony services, security in network protocols, Internet telephony signaling protocols, pervasive computing in the telecommunications domain, distributed systems programming, and programming languages. Vijay holds two patents and has six applications pending with the U.S. Patent Office. He is a member of the ACM and IEEE Computer Society.

Xian-He Sun is a professor of computer science at the Illinois Institute of Technology (IIT), the director of the Scalable Computing Software (SCS) laboratory at IIT, and a guest faculty in the Mathematics and Computer Science Division at the Argonne National Laboratory. He received a B.S. in Mathematics from Beijing Normal University, P.R. China, and holds his M.S. in Mathematics and Ph.D. in Computer Science from Michigan State University. Before joining IIT, he was a post-doctoral researcher at the Ames National Laboratory, a staff scientist at the ICASE, NASA Langley Research Center, an ASEE fellow at the Naval Research Laboratories, and a professor at the Louisiana State University–Baton Rouge. He is an internationally renowned researcher in scalable high performance computing and communication. He has published over 100 articles and papers in the area of high-performance and distributed computing and communication.

Acknowledgments

I am grateful to the many people who have made this book possible. It always goes without saying that the contributions to complete a project of this nature come from many sources. I am also indebted to Bell Laboratories, Lucent Technologies, Inc. for providing an atmosphere that has supported this work. The students of Prof. Xian-He Sun's Scalable Computer Systems Laboratory aided in many aspects of the work described in this book. Specifically, I thank Charles Jodel for spending his Thanksgiving transcribing portions of this publication.

VKG

Like many of my peers growing up in China during the Cultural Revolution, I did not receive a good education as a youth and began working full time at age 15. From taking the college entry exam to arriving in America for advanced study, from locating my first research position to receiving my first research grant, I have had many failures and few successes. I am heartily grateful to the many people who have helped me to make the few successes that define who I am today.

XHS

List of Abbreviations and Symbols

Abbreviation	Definition
λ	Arrival rate per time unit
μ	Service rate per time unit
ρ	Traffic intensity
$\varnothing(x)$	Call Model Mapping function
P_α	Policy tuple
ϕ	Constraint on P_α
$B(c, \rho)$	Erlang-B blocking probability
2G	Second Generation network
2.5G	Data enhanced Second Generation network
3G	Third Generation network
3GPP	Third Generation Partnership Project
AC	Authentication Center
API	Application Programming Interface
ATM	Asynchronous Transfer Mode
B2BUA	Back-to-Back User Agent
BCSM	Basic Call State Model
BS	Base Station
CA	Certificate Authority
CGI	Common Gateway Interface
CMM/SS	Call Model Mapping with State Sharing
CO	Central Office
CORBA	Common Object Request Broker Architecture
CPL	Call Processing Language

CSN	Circuit Switched Network
CTI	Computer Telephony Integration
DNS	Domain Name Service
DP	Detection Point
DTMF	Dual Tone Multi-Frequency
EM	Event Manager
FE	Functional Entity
FEA	Functional Entity Action
FSM	Finite State Machine
FTP	File Transfer Protocol
GPS	Geographical Positioning System
HLR	Home Location Register
HTTP	Hypertext Transfer Protocol
IANA	Internet Assigned Numbers Authority
ICW	Internet Call Waiting
IETF	Internet Engineering Task Force
IF	Information Flow
IM	Instant Message (or Instant Messaging)
IN	Intelligent Network
INAP	Intelligent Network Application Part
IP	Internet Protocol, *also* Intelligent Peripheral
ISUP	ISDN User Part
ITU	International Telecommunication Union
ITU-T	International Telecommunication Union–Telecommunication Standardization Sector
JAIN	Java API for Integrated Networks
JCAT	Java Coordination and Transactions
JCC	Java Call Control
JTAPI	Java Telephony Application Programming Interface
MC	Message Center
MIME	Multipurpose Internet Mail Extensions
MSC	Mobile Switching Center
O_BCSM	Originating Basic Call State Model
OSA	Open Services Architecture
PDU	Protocol Data Unit
PE	Physical Entity
PIC	Point(s) in Call
PSTN	Public Switched Telephone Network
RFC	Request For Comment
RPC	Remote Procedure Call
RTP	Real-time Transport Protocol
SCEP	Service Creation Environment Point
SCP	Service Control Point

SDP	Session Description Protocol, *also* Service Data Point
SIB	Service Independent Building Block
SIP	Session Initiation Protocol
SME	Short Message Entity
SMP	Service Management Point
SMS	Short Message Service
SMTP	Simple Mail Transfer Protocol
SN	Service Node
SOAP	Simple Object Access Protocol
SS7	Signaling System Number 7
SSP	Service Switching Point
T_BCSM	Terminating Basic Call State Model
TCAP	Transaction Capabilities Part
TDM	Time Division Multiplexing
TINA	Telecommunications Information Networking Architecture
TLS	Transport Layer Security
UAC	User Agent Client
UAS	User Agent Server
UDDI	Universal Description, Discovery and Integration
URI	Uniform Resource Identifiers
VLR	Visitor Location Register
WIN	Wireless Intelligent Network
W-LAN	Wireless Local Area Network
WWW	World Wide Web
XML	eXtensible Markup Language

Chapter 1

Introduction

Service-oriented computing is the computing discipline that views services as the fundamental elements for developing applications and solutions [PAP03]. This book is about orchestrating services that execute across different networks and answering the challenges such an arrangement inevitably poses. Specifically, we explore service-oriented computing in the context of enabling cross-network services between two communication networks: the Internet and the public switched telephone network (PSTN). We use the term *PSTN* in this book to encompass two aspects of the switched telephone network: wireline networks and cellular networks. Unless specified otherwise, the term will refer to both aspects of the switched telephone network.

We define a service as a value-added functionality provided by network operators to network users. Thus, making or receiving a call is a PSTN service, as is call waiting and caller identification. Instant messaging (IM), presence, electronic mail, and the World Wide Web (WWW) are examples of Internet services. We are primarily concerned with two networks in use today: the Internet and the PSTN. To a great extent, these networks have been influenced by each other. For instance, since the early days of the Internet, the PSTN infrastructure (telephone lines) has been used to transport Internet traffic. Even today, Internet users routinely access the Internet through their phones. Conversely, the Internet is perfectly capable of digitizing voice traffic between two communicating users and transporting it as data packets. However, much more advanced interactions are possible between the two networks — interactions that go beyond one network using the other as mere transport. The association of the

PSTN and the Internet in this manner (i.e., one network using the other as transport) was simply a very early prerequisite of the more advanced interactions that are proposed in this book.

The general realm of this book lies in Internet telephony; however, Internet telephony subsumes a substantial body of knowledge and area of research. The most visible form of Internet telephony is Voice-over-IP (VoIP), which can be defined as the ability to packetize and transport predominantly voice, but generally any communication-based content, including video and facsimile, over the Internet instead of the PSTN. However, Internet telephony is more than VoIP; it also encompasses aspects of an enhanced communication experience using the services of a general-purpose network such as the Internet.

Internet telephony does not exist in a vacuum, it coexists with incumbent networks (PSTN) and technologies; as such, it must cooperate with them [FAY00, MES96, MES99, SCH98]. The cooperation extends across two planes: the transport plane (i.e., the protocols and procedures for digitizing and transporting voice as packets over an inherently best-effort delivery network) and the service plane (i.e., the protocols and procedures for executing services in a network). The work described in this book pertains exclusively to the service plane and is part of an overall research approach for enabling what we call *crossover services*, i.e., services where the intelligence to execute them is distributed in multiple networks [GUR03b]. Note that we do not consider digitizing a voice stream and transporting it as packets across the Internet a crossover service; this is not because it is simple (it most definitely is not), but rather because that sort of a service is better addressed in literature that deals with the transport plane. We are most interested in working at the service plane to examine the cross-pollination of ideas that results when events in one network are used as precursors for services in another network. These services, as is the intelligence to execute them, are distributed across network boundaries.

A crossover service occurs as a natural by-product of using two communication networks. As these networks continue to merge, it becomes imperative to share the services across the networks. In some cases, the service itself executes on the PSTN and needs to be accessed from an Internet endpoint; in other cases, the service executes on the Internet based on discrete events occurring in the PSTN. To the user participating in the service, the details of the residency and execution of the service are immaterial. The service is simply a value-added functionality provided by the underlying networks. Realizing crossover services is complex; both networks in question use dissimilar protocols, procedures, and architectures for service execution. Their differing views have to be reconciled to produce a working crossover service.

First, some background on Internet telephony is required to understand the critical role that services play.

1.1 The Evolution of Internet Telephony

The beginnings of Internet telephony can be traced to 1998. The Internet had by then already achieved widespread deployment. It had successfully moved from its roots in academia and commercial research labs to mainstream adoption. The two most recognizable facets of the Internet were electronic mail and the World Wide Web (WWW). With the advance of the Internet, academic research and commercial laboratories started to pay closer attention to digitizing voice and transporting it as discrete packets across the Internet.

To be sure, the idea of packetizing voice was not new. It has been a subject of research ever since packet-switched networks have been in existence [COH76, COH77, WEI83] and continues to be so [DAS03, GAR03, HOM03, JIA03, MAR03, TSE04]. What was new in 1998 were four things: First, the widespread availability of a global network in the form of the Internet ensured reachability among its participants. Second, computing power had matured to the point where it was feasible to encode and decode voice packets in real time, even in handheld devices. Third, the collective knowledge in the field of real-time transport of delay-sensitive data (like voice) was coalescing around a set of standards — Real-Time Transport Protocol (RTP) [SCH03], Session Description Protocol (SDP) [HAN98], International Telecommunication Union — Telecommunications Standardization Sector's (ITU-T) H.323 [ITU03], and Session Initiation Protocol (SIP) [ROS02] — that could be implemented by organizations other than telecommunication vendors. And finally, the Telecommunications Deregulation Act of 1996 created a level playing field by forcing the incumbent telephone service providers to share their equipment and network with upstarts. The combination of these four effects resulted in a gradual shift in telecommunications from the circuit-switched nature of the PSTN to the packet-switched nature of the Internet.

Early Internet telephony was characterized by emphasis on the media (voice in this case). Internet telephony was viewed as a means to get around paying the telecommunication operators money for using their networks (a practice called toll arbitrage). If, instead, people could use their personal computer to digitize voice and the Internet to packetize and transport it, they would not have to pay the telecommunication operators for the privilege of communicating with others. Toll arbitrage was a powerful motivator at the onset; many start-ups were funded to create dense port voice gateways that would convert circuit voice to packets, yet others were funded to demonstrate better ways of multiplexing more voice channels over a transport or to dream of a better codec.

However, this stage did not last for long. Incumbent telecommunication operators, sensing the threat, countered by lowering voice tariffs on local and long-distance calls. This continued to the point where the rates to set up a circuit call were about the same as those for an Internet telephony call. Because the quality of the voice was much better on the circuit-switched network than it was on an unmanaged and best-effort delivery network like the Internet, Internet telephony had to find a better answer than toll arbitrage. Thus, Internet telephony entered into its next (and current) shift: emphasis on services [GLI03].

The shift toward services was further accelerated by the advent of the Third-Generation Partnership Project (3GPP). The Third-Generation (3G) mobile network, when fully deployed, will be based on the Internet Protocol. With endpoints that are Internet connected and far more powerful than those of the current cellular[1] network (known as the Second-Generation (2G or 2.5G) network), 3G envisioned personalized telecommunication services for each individual. 3G also envisioned a fast data pipe (supporting speeds up to 384 kbps) between the network and the cellular phone that will enable the network to deliver video to the phone and allow the phone to send pictures and packetized voice or video content to the network for further distribution.

3G and wireless local area network (W-LAN) have simply cemented the future of services in Internet telephony. In fact, as the means by which users communicate — wireline PSTN, cellular PSTN, 3G, wireline Internet, wireless Internet — proliferate, the need for crossover services will increase [LOW97]. The plurality of networks means that one network is not going to dominate in the future; some networks specialize in certain services, while others are expensive to replace entirely. Regardless of the network employed, services will play an important role [GLI03].

1.2 Problem Statement

The plurality of the communications networks and the emphasis on services frame the problem addressed by this book. Put succinctly, the problem is how best to foster service accessibility of existing services across different networks and how best to foster service innovation by

---

[1] In existing literature, the terms *cellular* and *wireless* are used synonymously to refer to the cellular PSTN; however, with the advent of the wireless Internet (in the form of the IEEE 802.11 wireless network or wireless fidelity (Wi-Fi)), there is a chance for ambiguity if these terms are not qualified further. Hence, this book uses the term *cellular* to refer to the cellular PSTN and *wireless* to refer to the wireless Internet.

utilizing the capabilities inherent in each network that participates in the service invocation. The problem is further exacerbated in two ways: First, because the state of the service will be shared across at least two networks, synchronization of the entities that participate in the service becomes paramount. As is the case with any distributed system, synchronizing the attendant entities is of utmost concern to yield a predictable system. Second, the signaling protocols and the finite-state machines used to model the progress of a service (or a call) will vary among networks. Thus, some order must be imposed such that all entities participating in the service view the progress of the service in a uniform manner.

The Internet and PSTN have very dissimilar ideas on how services are executed. The Internet espouses control of the service at the edge of the network, while the PSTN is most comfortable with centralized control of services. Messerschmitt [MES96] predicted that when these networks converge, one of the main challenges will be "interoperability across heterogeneous terminals and transport environments, and integration of heterogeneous services and applications within shared-resource environments."

1.3 Solutions

As discussed, services are viewed as the most important ingredient in Internet telephony. As Internet telephony progresses, services will follow three stages. In the first stage, users of Internet telephony will simply expect that the PSTN services they are accustomed to will be available in Internet telephony endpoints. The second stage will be characterized by cross-pollination of service ideas between the networks. More specifically, Internet-type services will merge with PSTN-type services. As a quick example, presence is an Internet service that currently is defined for a user using a Internet device (e.g., a personal computer) to log into a presence server. The presence server subsequently disseminates the presence information to interested parties. The act of logging in triggers the presence service, i.e., the user is present at a particular place. Similarly, the act of picking up a telephone connected to the PSTN can trigger the presence service, i.e., a user is present at home or at work when he interacts with that device. The cross-pollination of ideas will be even more important as the number of networks over which a user communicates increases.

The third and final stage of service evolution will be characterized by the applicability of services resulting from the cross-pollination of ideas in new disciplines, such as pervasive or ubiquitous computing. Accordingly, the solutions are organized around the service stages just outlined.

1.3.1 The First Stage: Accessing Native PSTN Services from Internet Telephony Endpoints

Although much work has been published on call establishment across the PSTN and Internet [CAM02, LEN01b, VEM02], much less progress has been made on how the signaling for services can be effectively carried out. The services that a telephone user is accustomed to reside and execute on the PSTN. Such services include call waiting, 800-number translation (for instance, translating a nationwide 1-800-GET-A-PIE number into the local number of a nearby pizzeria, pertinent to the area from which the call originated), call blocking (parental control of outgoing 900-number calls), etc. These services need to be provided from the newer Internet endpoints as well, preferably without rewriting the entire set of services that already exist and execute on the PSTN. We propose a technique termed call model mapping with state sharing (CMM/SS), which demonstrates the feasibility of providing native PSTN services from Internet endpoints. This technique is general enough to be applicable to a variety of Internet signaling protocols. We will describe CMM/SS in detail and demonstrate its feasibility by an implementation. We will also present performance characterization of services executing natively in the PSTN versus services executed through the CMM/SS technique.

1.3.2 The Second Stage: PSTN Events as a Precursor for Internet Services

The PSTN is a veritable storehouse of interesting events, such as the arrival of a call, the initialization of a call (making a call), analyzing dialed digits, location updates in the cellular network, cellular endpoints registering and de-registering themselves, and many more. All these events can be harnessed to provide services on the Internet. To do so, an ontology is required to allow an Internet host to communicate with the PSTN entities generating these events. Issues such as quantifying the events in a uniform manner, representing them in a protocol understood by both the Internet and PSTN entities, synchronizing the Internet and PSTN entities, and privacy in and security of such a system all become important research challenges. We propose an architecture and discuss an implementation that allows Internet hosts to leverage the events in the PSTN for service execution in the Internet. The architecture addresses the problems outlined above and is general enough to be applicable to the cellular as well as wireline aspects of the PSTN. We also establish a taxonomy of services that can be executed based on PSTN events. Establishing a classification for such services is important so that implementers can quickly identify various techniques for rapid implementation.

1.3.3 The Third Stage: Pervasive Computing and Telecommunication Services

"The most profound technologies are those that disappear. They weave themselves into the fabric of everyday life until they are indistinguishable from it" [WEI91, p. 91]. This was Mark Weiser's vision of pervasive computing. A case can be made that the telecommunication network has already woven itself into the fabric of everyday life; the Internet is doing so now. The services provided by the two networks as they converge lead to the creation of a telecommunication smart space [SAT02]. A smart space is an aggregate environment composed of two or more previously disjoint domains. As a final contribution, we demonstrate how the PSTN and the Internet cooperate to create a smart space in the telecommunications domain. This smart space leads to many innovative services and service ideas that build upon the strengths of the individual networks involved in the service. Indeed, in the Internet and PSTN convergence, [MES99] makes a case for a framework that builds upon the strengths of both networks. The third stage in the service evolution will be characterized by many such specialized frameworks; the application of pervasive computing to telecommunication services is one such framework.

In this book, we propose strategies and techniques to make crossover services a reality and the application of such services to related disciplines, such as pervasive or ubiquitous computing. These strategies begin with the idea of service-oriented computing, articulated at the beginning of this chapter — namely, that services are the fundamental elements for developing applications and solutions. The first step to a general service-oriented architecture (defined later) in the telecommunications domain lies in the application of service-oriented computing precepts to this domain.

Chapter 2

Internet Telephony: The Evolution to a Service-Oriented Architecture

The field of telecommunications has evolved to resemble the distributed computing domain, where general-purpose computers communicate over a common network. This evolution has culminated in the Third-Generation Internet Multimedia Subsystem (3G IMS) Architecture. In the computing domain, the Web Services Architecture or, in general, the Service-Oriented Architecture (SOA) is the modern trend in distributed computing today. Both the IMS and the Web Services Architecture provide services built on common, standardized, and well-known protocols. In this chapter, we present our views on the continuing evolution of these two architectures as the boundary between telecommunications and computing services continues to blur. We provide a high-level architectural overview of both the telecommunications and Internet networks to provide a context for the requirements we derive for a telecommunications SOA. Detailed architectures of both the networks will be provided in Chapter 3.

2.1 Introduction

Until recently, the line between the telecommunications network and the Internet was well demarcated. The former was a special circuit-switched network, tuned to transporting one media: voice. Over the years, it had also evolved to provide voice-related services to its users — colloquially known as subscribers — such as call forwarding, call waiting, and other services. In the telecommunications network, intelligence was concentrated in the core of the network, with the edges (phones) being very simplistic. The Internet, on the other hand, resided at the opposite spectrum from the telecommunications network. It was designed as a packet-switched network that would transport any type of media — voice, video, gaming, text — in a packet. The core of the Internet was relatively simple and stateless — it only performed the routing of packets; the intelligence resided at the edges of the network in the form of powerful general-purpose computers [KEM04].

More recently, the established lines between the Internet and the telecommunications network have started to blur. Today's 3G cellular phone is capable of providing many Internet services: e-mail, presence, Web browsing, and instant messaging, to name a few. On the other hand, transporting voice, which was once thought to be the domain of the traditional telephone network, is now done by the Internet. Two recent advances in technology have aided in this shift. The first advance is related to the form factor of what constitutes a computing device. Moore's law and other advances have continued to shrink hardware components to the point that a fairly sophisticated computer can be embedded in a handheld telephone. Second, the advances in the field of networking have made the Internet faster and more pervasive than ever. These two advances have caused each of the networks to steadily encroach on the principles held dear by the other.

In the telecommunications network, we note that the intelligence is being pushed to the edges. Witness the rise of the cellular network with much more sophisticated endpoints than the traditional telephone network. For its part, the Internet has started to adopt certain services in the core of the network, especially those pertaining to federated computing and quality of service. Two good examples of this are grid computing and Internet telephony. According to one definition [FOS], grid computing is comprised of systems that "use open, general purpose protocols to federate distributed resources and to deliver better-than-best-effort quality of service." To do this, some intelligence in the core network is necessary. Internet telephony also depends on some intelligence in the network to maintain the quality of service of a voice or video session. A "flow control" label maintained in a packet aids the Internet routers in associating incoming packets with an existing stream and providing the guaranteed quality of service to that stream. Admittedly, this is "soft state" [KEM04] in the network when compared to the state of a call maintained by a traditional telephony switch, but it is state nonetheless.

The telecommunications and Internet architectures continue to evolve toward a single network, on both the physical and transport layers, as well as the service layer. The first step in the evolution was the merging of the physical and transport layers. (Voice is packetized and transported on the Internet. In an ironic twist, the telephone network is able to provide a Digital Subscriber Loop (DSL) broadband connection that can be used to transport voice packets, thus bypassing the telephone network for transporting voice. But the larger point is that at the physical and transport layers, the Internet may have usurped the circuit-switched telecommunications network.)

The second and probably more important step is the evolution of the services layer. Both the telecommunications network and the Internet have their own service architecture. As these networks merge, one model of services — the computing style of services or the telephony style — will prevail. Which service architecture is better suited for the future? Although we do not have an answer to this question, we do provide an analysis of the service architectures of telecommunications and the Internet and draw parallels between them to extrapolate some attributes that may be present in an overall architecture in the future.

2.2 Service Architecture for Traditional Telephone Network

The service architecture for the traditional telephone network (wireline and wireless) is defined around the Intelligent Network (IN) [FAY96]. IN is a conceptual architecture that separates the call control from the service execution. Figure 2.1 shows a simplified IN architecture. (In reality, there are more IN entities than depicted in Figure 2.1, but for our discussion, the ones depicted in the figure suffice.) Subscribers use telephonic devices that are connected to a telephone switch called the Service Switching Point (SSP). An SSP, in turn, is connected to yet other IN entities via a packet network called Signaling System 7 (SS7). The most important IN entity is the Service Control Point (SCP), which is added to the call by the SSP. An SCP is a general-purpose computer that hosts and executes the service logic for a subscriber. The service logic can invite other IN devices into the call; for instance, if a service requires the caller to interact with a voice response system, an Intelligent Peripheral is dynamically added to the call.

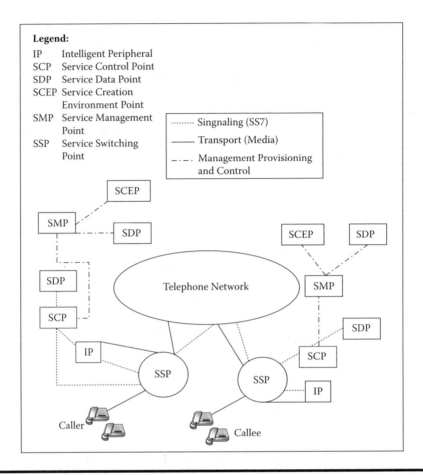

Figure 2.1   Traditional telephone network architecture.

The service logic can also access data pertinent to each subscriber stored in a specialized database called the Service Data Point (SDP).

When a call is originated on the telephone network, the caller's SSP arranges for the SCP to be brought into the call. The SCP then executes the service logic, depending on the services subscribed to by the caller. This process is repeated on the callee's side as well; the callee's SSP on receiving a call setup request arranges for an SCP to be brought into the call, and so on.

Services themselves are created in a general-purpose computer called the Service Creation Environment Point (SCEP). A service in IN is created by chaining reusable components called Service Independent Building Blocks (SIBs); many well-known SIBs exist, such as number analysis or adding new devices into a call. Service logic programmers employ a SIB palette to drag and drop individual SIBs to compose a service. Once a service is thus created, it is deployed at the SCP using the Service Management Point (SMP), which is yet another general-purpose computer through which service management and provisioning are performed.

The cellular telephone network also uses IN to create and deploy services. The process of executing a service in the cellular network is similar to its wireline equivalent; the crucial difference is that in a cellular network, there are more entities involved in providing a service. A set of databases — home location register and visitor location register — track the subscriber and store the services associated with the subscriber. Authentication servers authenticate and authorize a cellular endpoint, and other infrastructure (base stations, mobile switching centers) provide radio access networks and the capability to connect to other cellular subscribers or to the wireline telephone network.

More information on IN is provided in [FAY96].

2.3 Internet Services Architecture

At the onset, the Internet supported predominantly client–server type of service architectures, wherein a client made a request of a server and the server provided the service. The service could be as simple as a one-time generated piece of information (for instance, ldap, ntp, etc.), or it could be more complex, like a ftp or a telnet session. The next step to the client–server architectures was the move toward distributed objects.

Distributed object frameworks — like Distributed Component Object Model (DCOM), Common Object Request Broker Architecture (CORBA), and Remote Method Invocation (RMI) — hid the complexities of client–server interactions in well-defined application programming interfaces (APIs). Programmers no longer needed to worry about Internet addresses and port numbers to access servers, or even how to structure the protocol data unit for a service. Precompilers read an interface definition file and generated the corresponding server skeletons and client stubs. The application programmer had to only fill in the application-specific logic in the generated code. Over the years, distributed object frameworks have evolved to provide message reliability and transactional guarantees.

The next step beyond distributed objects is service-oriented computing (SOC) and service-oriented architectures (SOAs), as exemplified by Web services. Papazoglou and Georgakapoulous [PAP03a] describe SOC best as a "computing paradigm that utilizes services as fundamental elements for developing applications." They go on to provide the following description of SOA: "Basic services, their description, and basic operations (publication, discovery, selection, and binding) that produce or utilize such descriptions constitute the SOA foundation."

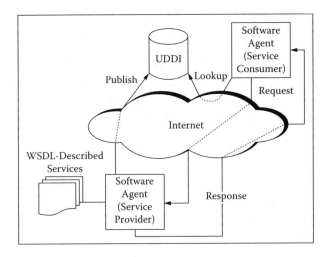

Figure 2.2   Web Services Architecture.

SOAs are characterized by loose coupling between the interacting software agents. The software agents exchange well-defined messages to execute services. The format and semantics of these messages are collectively defined by Web services. Web services transport eXtensible Markup Language (XML) documents between software agents using standard Internet protocols. Software agents that send and receive these documents implement the semantics of a particular service.

Web services are described using the Web Services Definition Language (WSDL), which is an XML format for describing network services as a set of endpoints operating on messages [CHR01]. WSDL describes all information pertinent to a Web service: its name, transport binding (using Simple Object Access Protocol (SOAP), Hypertext Transfer Protocol (HTTP), or Multipurpose Internet Mail Extension (MIME)), and a set of operations supported by the service. Once described as such, services are stored in a Universal Description, Discovery, and Integration (UDDI) registry. Clients can query the registry to get a listing of all matching services. Figure 2.2 contains an overview of this process. A software agent (service producer) describes the services it supports through WSDL and registers them with a UDDI registry. At some later point, another software agent (service consumer) looks up the service in the UDDI and contacts the service producer directly to execute the service.

2.4 Requirements of a Telecommunications SOA

Unarguably, the intelligence in the telecommunications networks continues to move out to the edges. Early analog phone systems were completely centralized. Call processing and service execution were often intertwined and occurred on centralized platforms owned by the telephone company, and also often used networks owned by the same telephone company. Digital switching systems coincided with the separation of call signaling from service execution (the IN concept). The intelligence was more distributed here and, in fact, it was not strictly necessary that the service execution platform be owned by the same provider that owned the switch (although this was, by far, the most common deployment scenario).

The advent of the mobile cellular network furthered the move of intelligence to the edges. The first-generation (1G) cellular network was an analog circuit-switched system. Mobile handsets were bulky, voice quality was poor, and security was nonexistent. 2G networks improved on the disadvantages and provided additional data services, like Short Message Service (SMS). 2.5G is an intermediate step toward 3G, utilizing Internet protocols and packet switching in portions of the cellular network. 4G networks are a step beyond 3G, providing data transmission speed equivalent to a local area network and more personalized services for its subscribers. As can be observed, in successive generations starting from 2.5G, the Internet Protocol plays an increasingly bigger role, and the mantra of the Internet Protocol is that intelligence resides at the edges [KEM04]. Thus, if the underlying telecommunication technology is moving toward the Internet Protocol, there appears to be some promise in the argument that the telecommunications service architecture will also evolve toward a computing-based one. Initial stages of this progression are characterized by the crossover services model, whereby both networks are cooperating to execute the services [GUR04a, GUR05a]. The final stages will witness the wholesale move of telecommunication services to the Internet, thus establishing a telecommunications SOA.

The as yet unanswered question is: What shape will this new architecture take? Let us first examine the similarities and differences between the service models of the two networks, and then attempt to tease the requirements of a telecommunications SOA.

The biggest similarity between the service models is that they both use message passing as an underlying mechanism for service execution. In the telecommunications network, an SSP sends a well-formatted message — a request — to the SCP, which in turn sends a well-formatted response back. This interaction matches the Web services description of exchanging structured information in a distributed environment. Thus, at a very high level, both networks contain primitives to name resources and route messages between the resources.

Closely related to message passing is the desire to use standardized protocols to transport these messages. Within their respective domains, both Web services and telecommunications services use standardized protocols for message passing. Web services use XML, SOAP, UDDI, and WSDL, whereas telecommunication services use protocols such as SS7 [RUS02] and Intelligent Network Application Part (INAP) [FAY96].

A second similarity lies in the manner services are defined within the computing discipline. In computing, a service can be described as an autonomous, platform-independent computational element. This is also true of telecommunication services. The autonomous, platform-independent computing element in telecommunications services is called a SIB (introduced in Section 2.2). Application programmers in telecommunications typically use a palette of well-known SIBs; creating a service is as simple as dragging a SIB and connecting it to other SIBs through edges that represent the actions and events (of course, the service has to be compiled and tested before being widely deployed). A SIB, much like a computing service interface, absolves the programmer from knowing the internal protocol details [WIL97].

A third similarity is the need for orchestration. An example will illustrate: Consider a stock price Web service that returns the stock price of a certain company. In and of itself, it is a useful service, but when it is coupled with another service — call this the stock-purchasing service — that purchases the stock when it falls to a certain price range, the utility of the stock price service is increased. The need for orchestration arises because some intelligence is required to trigger the stock-purchasing service when the price falls below a certain watermark. The business community, lead by IBM, Microsoft, and others, is coalescing around an orchestration language called Business Process Execution Language (BPEL). Other domain-specific orchestration languages and coordination strategies will undoubtedly emerge as the Web Services Architecture matures.

In telecommunication services, orchestration is already present in the form of a call controller. A call controller can be thought of as a service that is instantiated when a caller initiates a call, or it can be instantiated based on other service logic (i.e., at 3:00 P.M., start a conference call between the following people...). The call controller is the intelligent centralized entity that controls all aspects of the call: it knows the number of parties in a call, the duration of the call, billing information of the call, and so on. As an advanced example of coordination in telecommunication services, imagine that there are two discrete services: a presence service that determines where a subscriber is present (home, in transit, work) and an instant message service that can communicate with the subscriber. The call controller can use the presence service to locate the called party, and it could use the instant message service to first send an IM to the called party and explicitly ask its disposition before allowing the called party's phone to ring.

Having examined the similarities, we now enunciate the differences. The first and most apparent difference between the two service architectures lies in the realm of discovery, selection, and binding of services. The Web Service Architecture includes WSDL and UDDI to aid in this effort; an equivalent mechanism in telecommunications services is missing. One possible reason for this is that in the telecommunication domain, the service vendor and the service provider typically have preexisting business relationships, thus alleviating the need for service discovery and binding. This is not the case for Web services. A Web service written by an arbitrary service vendor should, at least in spirit, work across the platforms of multiple service providers. This process automatically results in the best-of-breed services percolating to the top of the list. There is no equivalent for discovery, selection, and binding in the telecommunications network. Preexisting business relationships between the service provider and the network vendor have alleviated the need for this (traditionally, the service providers have been the equipment vendors, who create and sell services directly to the network providers, who in turn bill their subscribers for these services). *It thus follows that some manner of a discovery, selection, and binding mechanism based on open and standardized protocols should be a requirement of a telecommunications SOA.*

A second difference lies in the real-time nature of telecommunication services versus Web services. The latter are mostly data driven, i.e., move information from point A to point B. These services can withstand delay and are generally more forgiving in the face of a best-effort network like the Internet. Web services are also characterized by a single protocol for communications (SOAP) and data representation (XML). Telecommunication services, on the other hand, are both data driven and media driven (e.g., voice, video, gaming, and presence). They do not suffer delay in a graceful manner. This is true at both the media layer (delay in terms of lost packets introduces jitter and an inferior audio/visual experience) and the services layer (delay in accessing a database may prohibit the productive execution of a service). They also tend to be driven by a multiplicity of protocols: Session Initiation Protocol (SIP), SDP, Real-Time Transport Protocol (RTP), and Real Time Control Protocol (RTCP), to name a few. *Thus, a second requirement of the telecommunications SOA would be support for real-time delivery of information across a multiplicity of protocols.*

Another difference is in the security infrastructure of the networks. In the telecommunications network, security is addressed at two different layers. At the physical layer, the telephone lines and equipment (switches, base stations) and related databases are located in secure facilities. It is hard, although not impossible, to tap into a telecommunication line to eavesdrop on a call. If this itself was not a deterrent, the legal system has evolved to protect the conversation occurring across a phone line and levy fines for malicious access and disruption to the telephone service. The Internet, on the other hand, is a wide-open system by design. Tapping into it is no more work than enabling an Internet Protocol (IP) transceiver to go into promiscuous mode and sniff all the packets crossing the transport. The same level of legal scrutiny afforded to the telephone system has not yet made its way to the Internet. Another rather subtle reason for more security in the telecommunications system has been how its protocols have evolved. Telephone networks have traditionally been environments where the inner workings of the protocols and services, although not entirely secret, were not subject to as much public access and scrutiny as Internet protocols have been. Consequently, the number of people attempting to understand and implement these protocols, and indeed to mount malicious attacks, has been much less than the Internet equivalent. *Thus, we derive a third requirement of the telecommunications SOA: to provide security in a manner conducive to the Internet.*

Closely related to security is authentication. When a telephone subscriber makes a call, the network authoritatively knows the identity of the subscriber (because in the telecommunications network, the service provider owns the network and the subscriber data). Consequently, subscribers of the telecommunications network have learned to trust the content that flows over the telephone network. Services provided by the telephone network — call forwarding, caller ID, etc. — need not be further authorized or authenticated. The telephone service provider is the final arbiter on the authenticity of the service on that particular network. This is not the case on the Internet, where the provider of the transport (the network) is distinct from the provider of the content. The authentication problem on the Internet is made worse by the ease of mining new identities and the lack of a central trusted arbiter that can be used to rendezvous two previously unknown users. *Some means of authoritatively authenticating the communicating user agents is a final requirement for a telecommunications SOA.*

2.5 Conclusion

Our vision of a telecommunications SOA is that it will be an open, federated, secure architecture that allows the subscribers to choose the best-of-breed services from competing service providers and have these services work cooperatively.

The telecommunications SOA has to be federated. Services on it will require aggregation across autonomous boundaries. Take, for instance, using the presence service as a prerequisite to establishing a voice or video session between two participants. The presence state of the participants may be distributed across multiple devices (cellular phones, laptop computers, personal digital assistants, desktop computers) and service providers (Yahoo!, AOL, MSN, enterprise networks). Intelligence will be required to aggregate the presence and availability of a subscriber across states maintained in different autonomous systems. We note that grid computing, with its notion of virtual organizations, is well suited to solve portions of this puzzle. Further research is needed on the application of grid computing to telecommunications SOA.

The telecommunications SOA has to be built on open protocols. Not only will this aid in communications between federations, but also it will allow subscribers to describe their preferences and devices to enunciate their capabilities and store all of these in a secure repository accessible by trusted services. Open protocols will also allow devices to negotiate and prioritize services among themselves. In short, a robust discovery, selection, and binding mechanism for services in the telecommunications SOA will be another area of active research.

The telecommunications SOA has to be built with pervasive security. Devices must not be allowed to update a repository unless they can prove their authenticity. A subscriber must not be allowed to set up a session with another unless he can prove his veracity. Authentication and data integrity are that much more important in a telecommunications SOA because of the need for privacy that we accord to our communication needs.

And finally, the telecommunications SOA must support a multiplicity of service providers. In the Internet domain, innovation is played out as various providers — AOL, Google, Yahoo! — attempt to make themselves the preferred portal for a Web surfer by providing attractive services. Google, in particular, has played this card well. A Web surfer can use Google not only for browsing realty listings, but also to contact the listing realtor using Google Talk, take a virtual tour of the area in which the property is located using Google Earth, and then find directions to the property using Google Maps. Further innovation in the telecommunications domain will occur when multiple service providers vie for the attention of a subscriber. Providing extensible SOA frameworks for telecommunication services continues to be another area of fruitful research.

The evolution of a telecommunications SOA has just begun. The rest of this book provides an early glimpse into this evolution.

Chapter 3

Background: Providing Telephony Service

The principles of the public switched telephone network (PSTN) and the Internet are diametrically opposite each other. The former is a special-purpose network built to transport one communication aspect: voice. The endpoints are simple (a phone with a 12-button keypad and limited display capabilities, if any) and the intelligence associated with routing voice circuits and executing services is concentrated in the core of the network, where expensive computers called switches reside. The Internet, on the other hand, is a general-purpose network built to transport any media — voice, video, data, — using a best-effort delivery mechanism. The core of the Internet is fairly simple and consists of special-purpose computers called routers that receive and forward a packet toward the next router, and so on, until the packet reaches its intended destination. The intelligence in the Internet is concentrated at the edges in the form of powerful desktop and laptop computers and personal digital assistants.

Our work builds on the strengths of both of these networks. To provide a backdrop for interpreting the rest of the book, this chapter presents relevant background on the existing telephony service architectures of the PSTN and the Internet.

3.1 Service Architecture for the Wireline Public Switched Telephone Network

The PSTN is the most ubiquitous network deployed in the world, enabling billions of people to communicate. Besides providing basic communication capabilities, the PSTN also includes a well-defined service layer called the Intelligent Network (IN). This section provides the reader the requisite background to understand the relationship of PSTN/IN to the work described in this book. An in-depth treatment of PSTN can be found in [THO00, RUS02]; a detailed description of the IN can be found in [FAY96].

3.1.1 General Architecture of the PSTN

Figure 3.1 depicts the general architecture of the PSTN.

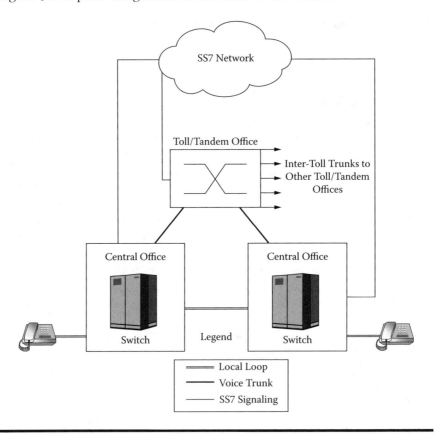

Figure 3.1   A high-level PSTN architecture.

Telephone users (also called end users or subscribers), in either homes or offices, connect to the telephone system through phones on their office desks or in their homes. Telephone traffic from end users terminates at a central office (CO) through a pair of wires (or four wires) called the local loop or the subscriber loop. A CO is owned by a telecommunications service provider responsible for providing service to a certain geographic area. Hundreds of COs may be installed in a metropolitan area. Telephone traffic from the COs is generally aggregated into trunks[1] and distributed to other offices.

Each CO contains one or more specialized computers called a digital switch, or simply a switch. The switch contains special-purpose hardware and software with stringent requirements on availability and fault tolerance. The switch is the brain of the PSTN; it shuttles the telephone network traffic between other switches and provides services to the end users. The CO further distributes the traffic. If the traffic is destined for a called party on the same switch, it does not go outside the CO; otherwise, it is sent to a toll/tandem office, which contains yet another special-purpose switch called a tandem switch — used to route traffic between COs. Because it is not physically possible to connect each switch to every other switch in the universe, a trunk from a CO connects to the tandem switch. The tandem switch connects to still other tandem switches, and so on, until it is possible for one CO switch to reach other CO switches through a tandem switch mesh.

A salient point about the PSTN is that the network used to route the media stream between switches is different from the network used to route signaling messages. Signaling messages between switches are routed over a packet-based network called Signaling System 7 (SS7). Communicating switches exchange SS7 packets to set up a call by allocating media resources end to end. Once the media resources have been allocated and the call has been set up, the voice flows over trunks between *each* intervening switch.

SS7 is a protocol stack consisting of four layers. The lower layers provide network connectivity and routing functions, while the topmost layer provides application-specific support. For telephony call setup and teardown, the topmost layer defines a format called ISDN User Part (ISUP). ISUP is the protocol used to set up and tear down telephone calls. The topmost layer also defines other application-specific data formats, as will be discussed later.

---

[1] In a communication network, a trunk is defined as a single transmission channel between two switches. Think of it as a wire that connects two switches and carries data between them.

The discussion on the PSTN thus far has not included any details on how services (besides voice transport) were provided to end users. In the early days of digital switching (circa 1986), very few services besides making (or receiving) a phone call were provided by the PSTN to end users. Line-oriented services such as call waiting and call forwarding were supported, but for the most part, the switches had the function of moving large amounts of telephone traffic efficiently among metropolitan areas. However, as computing power became more affordable and new technologies such as databases became commercially successful, the PSTN operators started toying with the idea of providing services like calling card and 800-number lookup services to end users. These services were executed on general-purpose computers with information stored in centralized databases. The centralized approach allowed the introduction of some services that would otherwise be impractical due to the complexity of managing large amounts of volatile data at every switch [BER92, FAY96]. However, the number of services was (and still is) limited and usually tied to the vendor of a particular switch, making it impossible to run the service in another vendor's switch.

Out of the need for a standardized service creation mechanism that would be vendor agnostic and provide primitives to create many new and exciting services faster than was currently possible was born the Intelligent Network (IN).

3.1.2 The Intelligent Network

The IN is an architectural concept; it provides for real-time execution of network services and customer applications in a distributed environment consisting of interconnected computers and switching systems [ITU92a, FAY96]. Until the advent of the IN, services were intimately tied to the switches and were not interoperable across vendor boundaries. The IN decoupled and distributed the call control and service execution to separate network elements; call control took place on switches and the service execution on general-purpose computers having access to fast databases, and on specialized devices to play announcements, collect digits, bridge calls, provide conferencing, etc. All connected to the switches through dedicated signaling links. The IN standardized the communication interfaces between a switch and a service platform as well as the service creation and management between them. The decoupling, distribution of functions, and standardization efforts had a great effect on how services were created and deployed on the PSTN. Services were now independent of the switch; they could be specified and implemented much faster and cheaper than before.

The IN is currently the *de facto* service architecture for the PSTN; its principles of distributing the intelligence among PSTN entities for service execution are very applicable to Internet telephony [FIN02]. The IN architecture has spawned a sizable number of research efforts [BAR93, CHI00, GBA99, LEN99, LIC01, PET00] and commercial interests — Sun Microsystem's Java Application Programming Interfaces (APIs) for Integrated Networks (JAIN)[2] [SUN05] and many industry consortia, primary among them Parlay/Open Services Architecture (OSA) [PAR05] and Telecommunications Information Networking Architecture (TINA) [TEL05]. The precepts of the IN architecture have been enormously influential to the general area of Internet telephony services.

3.1.3 The IN Conceptual Model

The International Telecommunication Union — Telecommunications Standardization Sector's (ITU-T) Recommendation Q.1201 [ITU92a] describes an IN conceptual model as a "framework for the design and description of the IN architecture." This conceptual model is then realized through a set of software protocols, finite-state machines, and associated hardware into a concrete IN architecture. The IN conceptual model has four layers, or planes. Each plane introduces an abstract view of the network entities, which is further made tangible in the plane below it. Starting from the top, these are the *service* plane, the *global functional* plane, the *distributed functional* plane, and the *physical* plane. Figure 3.2 depicts this hierarchy.

3.1.3.1 Service Plane

The service plane represents the designer's viewpoint of how a service should work. At this plane, services are described in terms of service features. A service feature is a service-independent aspect that describes one particular service but may be applicable to other services as well. An example provides more insight: A call-queuing feature describes the behavior of a call arriving in the network and required queuing if all lines that can service the call are busy. Incoming calls are queued and serviced on a first-come first-served basis as soon as a line becomes available. This

---

[2] In the late 1990s, when Sun initially released the JAIN API, the acronyms originally expanded to Java APIs for Intelligent Networks. However, driven by the nascent Internet telephony movement, the need for programming telephony services was so great that JAIN expanded beyond its IN roots. Thus, besides APIs for IN, there are now JAIN APIs for Internet telephony signaling protocols like SIP and SDP, services like instant messaging (IM), and many others. A complete list of JAIN APIs is provided in [SUN05].

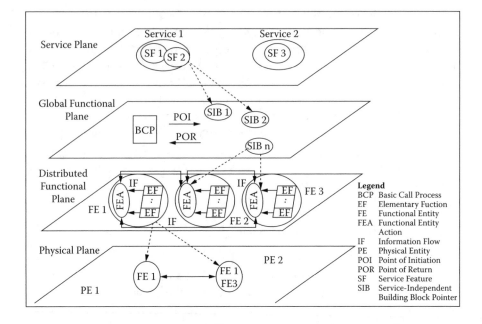

Figure 3.2   The IN conceptual model. (After Figure 20 in ITU-T, Principles of Intelligent Network Architecture, Recommendation Q.1201, Geneva, Switzerland, October 1992.)

call-queuing service feature can now be applied to the domain of a call center that has an 800 number for callers to dial in. When all agents are busy, the callers are queued; whenever an agent becomes available, the earliest caller in the queue is assigned to the agent. To observe service independence, note that the same call-queuing service feature can equally be applied to another domain: queuing incoming calls for the directory information (411) service. At this plane, services are described and composed in terms of independent service features.

3.1.3.2   Global Functional Plane

A service programmer observes the IN at this layer. The global functional plane provides a service programmer atomic building blocks from which to construct services. The service features of the service plane are mapped to atomic instructions called Service Independent Building Blocks (SIBs). The SIBs are reusable components and can be chained together to construct a service logic. The object handling the call runs a fixed finite-state machine called the Basic Call Process (BCP). When the BCP reaches a certain point (called point of initialization) that requires it to execute the service, further call execution is suspended and control is passed to the service logic. The service logic executes the SIBs and, upon completion, control is passed back to the BCP (point of return).

3.1.3.3 Distributed Functional Plane

This plane represents the view of a network designer. The entities in the network are viewed as a set of abstract models of software and hardware called functional entities (FEs). The FEs may perform atomic functional entity actions (FEAs) and, as a result of the FEAs, exchange messages — through remote procedure calls (RPCs), function calls, or electronic/fiber signals — called information flows (IFs). The SIBs of the global functional plane are realized by the sequence of FEAs in an FE. Certain FEs in this plane, especially those that model a switch setting up a call, play an important role in this book and will be examined more closely in later sections.

3.1.3.4 Physical Plane

This plane is of primary importance to network operators and equipment providers. The FEs of the distributed functional plane are mapped to physical entities (PEs) in this layer; for instance, the FE that controls the call will be realized as a switch, the FE that performs media services will be realized as a media server, and so on. PEs communicate with each other by exchanging protocol messages (which were represented as IFs in the distributed functional plane).

3.1.4 Physical Entities in an IN-Enabled Network

The IN conceptual model presented in the preceding section is, by design, an abstract model. When it is actually put into practice, the abstract entities are mapped into physical ones. This subsection explores the IN not from the conceptual or abstract model, but rather from a physical one involving the computers and other peripherals that actually render the network intelligent.

First, a high-level overview of how an IN-compliant service is executed will provide the relevant framework for the discussion that follows on the physical entities that make up the IN. An IN-compliant service is first constructed through an FE called the service creation environment function. This FE contains the programming environment, which includes the SIB that a programmer uses to construct an IN-compliant service. Once the service logic is created and tested, it is sent to another FE, the service management function. This FE deploys the service logic to the service execution FEs and allows for service customization.

An IN-enabled switch that is processing a call runs a fixed finite-state machine, the BCP. The BCP represents switch code and defines various control points, such as the point where the destination address has been received from the caller, or the point where the called user has answered the phone. When the BCP arrives at a specific control point, and certain prerequisites for executing a service are met, the BCP can trigger an RPC from the switch to a service execution platform.[3] The procedure call in the service execution platform runs the service; for example, the service may have been translating an 800 number by looking into a database. When the procedure call returns, the execution of BCP continues, using the translated number returned by the procedure call. Some services are far more complex than simple number lookup; for instance, a service could require the calling user to authenticate herself by typing in a passcode or uttering a password. Depending on the service logic, the service execution platform may involve other peripherals that provide functions to operate on the media to perform digit detection or analyze a spoken password.

With this high-level view in mind, Figure 3.3 outlines an IN-enabled network that contains the entities described in the service description above. This figure is an abridged version of Figure 1 in Q.1205 [ITU93], which defines all the FEs a PE may contain. For the discussion pertinent to this book, it is important to understand the most critical of the FEs and their corresponding association with a PE.

The PSTN augmented by the IN includes the following critical FEs (as shown in Figure 3.3).

3.1.4.1   Service Switching Point (SSP)

A switch that is capable of providing access to the IN capabilities is called an SSP; not all switches are so capable. The SSP provides users access to the telephone network through the local exchange. It acts as the first entry point into the IN; the detection capabilities of an SSP determine which of the subscribed IN services a user receives when he makes or receives a phone call.

---

[3] The RPC ultimately results in an on-the-wire protocol being sent from the switch to the service execution platform. In IN, this protocol is called Transaction Capabilities Application Part (TCAP) in the United States and Intelligent Network Application Part (INAP) in Europe. Both TCAP and INAP are application-level protocols (residing at the topmost layer of the SS7 protocol stack) and transported over the SS7 packet network.

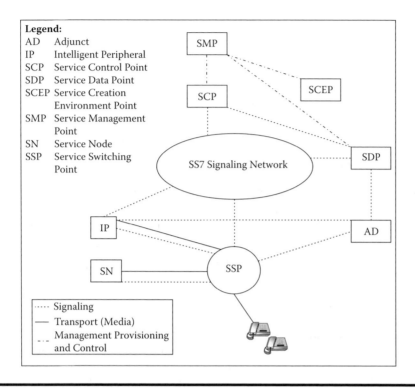

Figure 3.3 The PSTN augmented by the IN.

3.1.4.2 Service Data Point (SDP)

The SDP is a specialized database that contains customer data, which is accessed by the Service Control Point (discussed next) during the execution of an IN service. The SDP contains data that directly relates to the provision or operation of the IN services.

3.1.4.3 Service Control Point (SCP)

An SCP is a general-purpose computer connected to the SSP through the SS7 network. SCPs contain user programs and associated subscriber data (accessible through the SDP) that implement an IN service pertinent to a call. The SCP is brought into an IN service by the SSP; the SCP, in turn, can bring other IN entities into the service flow, if required. For example, when executing a prepaid card service, the SCP may use the services of an IN entity called an Intelligent Peripheral to perform digit collection or voice recognition.

3.1.4.4 Intelligent Peripheral

An intelligent peripheral is a specialized media resource server. It has physical access to the media stream of a phone call (see Figure 3.3); thus, it can provide media-related services such as voice announcements, speech recognition, dual tone multifrequency (DTMF) digit collection, audioconference bridging, tone generation, and text-to-speech synthesis.

3.1.4.5 Adjunct

An adjunct is functionally equivalent to an SCP, but is connected to a single SSP through a high-speed network (such as a local area network (LAN)), instead of an SS7 link. Certain IN features that require a fast response time between the IN service and the SSP can reside in the adjunct to take advantage of the high-speed connection between the SSP and the adjunct.

3.1.4.6 Service Node (SN)

An SN performs the role of an SCP and an intelligent peripheral. The IN services may reside and execute on the SN (as they do on the SCP) and, if they require DTMF or speech recognition, the SN itself can provide such functionalities (note from Figure 3.3 that much like the intelligent peripheral, the SN also has access to the media stream).

3.1.4.7 Service Creation Environment Point (SCEP)

An SCEP is a general-purpose computer where the IN services are programmed and tested before being deployed.

3.1.4.8 Service Management Point (SMP)

An SMP is a general-purpose computer through which service management and provisioning are performed. Services are loaded to the SCP for execution, and the data for the services is provisioned on the SDP.

3.1.5 The Basic Call State Machine, Points in Call, and Detection Points

The centerpiece of the IN conceptual model is a fixed finite-state machine called the Basic Call State Model (BCSM). A call state model, or call model for short, is a deterministic finite-state machine (FSM). It is represented as a digraph consisting of a set $V = (v_1, v_2, ..., v_{n-1}, v_n)$ of vertices and a set $E = (E_1, E_2, ..., E_{m-1}, E_m)$ of edges. A vertex is called a state and an edge is called a transition. There may be more than one transition leading into a state, and consequently, there may be more than one transition leading out of a state. Transitions corresponds to the events that occur during the processing of telephone calls, e.g., lifting the receiver to make a call, ringing, picking up the receiver to receive a call, dialing digits, etc.

Figure 3.4 depicts a sample call model with four states and eight transitions. v_0 is the initial state (also called the null state), and v_1, v_2, and v_3 are the three subsequent states. Transitions e_0, e_6, and e_7 lead into v_0, while e_1 leads out of it. In certain cases, transitions may lead back into the same state (as is the case with transition e_5).

Call models represented as FSMs serve two main purposes: first, they synchronize the various entities in the IN that provide services (SSP, SCP, intelligent peripheral, etc.), and second, they present a consistent view of a call to provide services. The latter deserves further explanation.

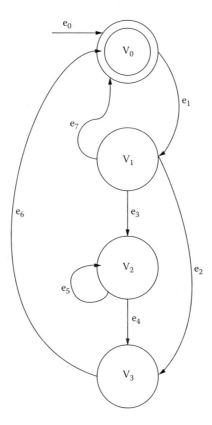

Figure 3.4 A call model represented as an FSM.

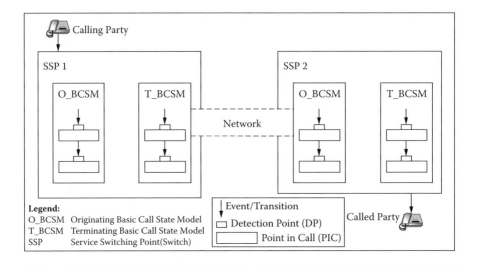

Figure 3.5   Originating and Terminating Basic Call Objects.

The BCSM is called a half-call model because it uses two halves to represent a call. Figure 3.5 contains a logical view of the call model. The half that originates a call is termed an Originating BCSM (O_BCSM); conversely, the half that terminates the call is referred to as a Terminating BCSM (T_BCSM). When a call is originated at an SSP, it initiates the O_BCSM and applies originating-side services[4] to the call. The SSP responsible for the ultimate recipient of the call initiates the T_BCSM and applies terminating-side services[5] to the call. Note that if the recipient of the call resides on the same SSP, the T_BCSM is attached to the recipient directly and terminating-side services are provided by the same SSP that provides originating-side services.

The BCSM, as stated previously, is a fixed finite-state machine. It has a certain number of states and a set of events that cause a change from one state to the next. The states are referred to as points in call (PICs), and the transitions between them are termed detection points (DPs). The PICs serve to synchronize the entities that participate in a call, while the DPs enable the service to be executed. Recall from the discussion at the beginning of Section 3.1.4 that when an IN-enabled switch reaches a certain control point, and certain prerequisites for executing a service are met, a call is suspended and an RPC is triggered from the switch to the SCP. The control point is a PIC, and the certain prerequisites are a DP being armed and satisfying some criteria associated with the DP.

---

[4] Originating-side services include 900-number blocking, 800-number translation, etc.
[5] Terminating-side services include caller identification, call waiting, etc.

DPs operate between the PICs; they delineate points in the model where call processing is suspended and the service execution platform is contacted to perform a service pertinent to that DP. Query messages are associated with each DP; when a query message is received at the service execution platform, the platform knows the exact state of the suspended call. A DP may be either armed or not armed; for the SSP to send a query message to the service execution platform, a DP must be armed and must meet certain trigger criteria. Examples of trigger criteria include bearer capability (DTMF or rotary dialing), presence of feature codes (for example, *70 in the United States is used to disable call waiting), or simply unconditional trigger (in which case, no other criteria are checked). A DP may be armed either statically or dynamically. Static DPs are armed through the SMP, as part of service provisioning. Once armed, a static DP remains so until explicitly disarmed by the SMP. A dynamically armed DP, as its name suggests, is armed on an as-needed basis by the SCP as it implements on the service logic. A dynamic DP remains armed as long as the SCP-to-SSP relationship persists (which is generally the duration of a call).

As far as call processing is concerned, either of the following two actions may be requested of the SSP when a DP is encountered:

1. The state of the call is encapsulated in a query message that requests further instructions from the SCP; the SSP suspends further call processing until a response is received.
2. The call processing continues normally and a notification of the event (the DP encountered) is sent to the SCP.

Accordingly, two attributes, R(equest) and N(otification), are defined for DPs, corresponding respectively to the two actions above.

3.1.6   The IN Capability Sets

The IN has been standardized incrementally, starting with a baseline set of services and associated call models and protocols called Capability Set (CS)-1, standardized in March 1993. CS-1 was intended to support primitive services only, i.e., those services that apply to only one party in a call. Sample services enabled by CS-1 included abbreviated dialing (dialing the last four or five digits to complete a phone call), call forwarding, and originating call screening.

CS-2 followed in September 1997, and in addition to support for CS-1 services, it contained more complex services, including support for personal mobility. Sample services supported by CS-2 included call waiting, multiparty call handling (call transfer, conference calling, etc.), and mobile registration/de-registration. CS-3 was released in December 1999; besides supporting CS-2, CS-3 also included support for IN/Internet interworking for the first time. In addition, CS-3 introduced other services, such as number portability, support for prepaid calling cards, and further work on user and device mobility. In August 2001, CS-4 was released, defining a further evolution of CS-3 services. CS-4 further cements the IN/Internet interworking on many fronts. It supports services such as Internet telephony (i.e., transport of voice over a packet network) and establishes the IN as an overlay service network common to all transport and signaling technologies. In fact, CS-4 uses the ideas developed in Chapters 5 and 6 of this book to interwork portions of the IN and the Internet (see Sections 6.1 and 6.4 of Q.1244 [ITU01], respectively).

3.1.7 Originating BCSM (O_BCSM)

PICs and DPs play an important role in the execution of the IN services and deserve more attention because the work described in this book makes considerable use of them. CS-1 and CS-2 were defined with their own call models, complete with their own PICs and DPs. Both call models are a subset of the IN BCSM defined in Q.1204 [ITU92b]. The BCSM defined in Q.1024 is independent of the capability sets; thus, it serves as a representative BCSM for the work described in this book. The state machine for O_BCSM of Q.1204 is provided in Figure 3.6. It contains 11 PICs and 21 DPs.

Each PIC has certain exit criteria defined in the form of a DP that is set, i.e., the DP is armed and the trigger conditions have been met. Some PICs have more than one exit criterion. PICs 2 through 9 have a common exit criterion in the form of DP 21 (Calling_ Party_Disconnect_and_Abandon) being set. This DP is set if the caller disconnects the call while the call is still active, or prematurely abandons further call processing (i.e., the caller hung up before call processing could be completed). PICs 7, 8, and 9 have another common exit criterion in the form of DP 18 (Mid_Call). This DP is used to implement mid-call services, as is the case when the call-waiting service plays an audible beep to the caller, resulting in the caller pressing the hook-flash to answer the incoming call, or if the caller subscribes to the three-way calling service, he or she may depress the hook-flash to add a third party to an existing call. Because exit through DP 21 and DP 18 is common to many PICs, the description of the PICs below will omit them and discuss other exit criteria in detail.

PIC 1 — O_NULL: This is a catchall PIC that absorbs exceptions resulting from processing the call (see transitions leading from DPs

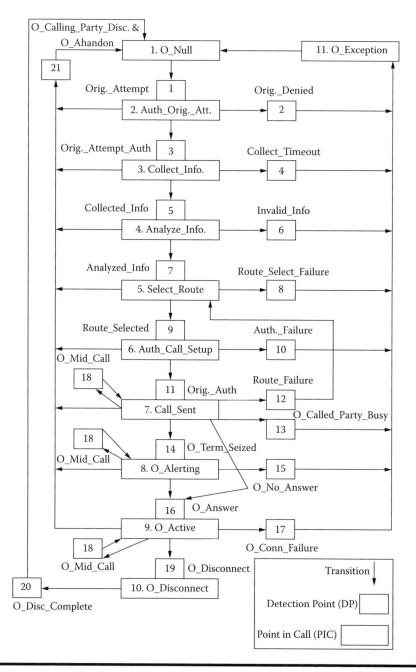

Figure 3.6 Example Originating Basic Call State Model. (After Figure A.2 in ITU-T, Intelligent Network Distributed Functional Plane Architecture, Recommendation Q.1204, Geneva, Switzerland, March 1993.)

20 and 21 and PIC 11 to this PIC in Figure 3.6). At this point, the call does not really exist. The only way to exit this PIC is through DP1, Origination_Attempt. If DP1 is set (armed and trigger conditions have been met, which will always be the case for DP1), control passes to the next PIC.

PIC 2 — Authorize_Origination_Attempt: At this point, the SSP has detected that someone wishes to place a call. Under some circumstances (e.g., use of the line is restricted to a certain time of the day), the SSP may not allow initiation of a call. Such services will be provided in this PIC. In addition to DP 21, PIC 2 has two exit means: through DP 3 (Origination_Attempt_Authorized) to PIC 3, or through DP 2 (Origination_Denied) to PIC 11. The processing of DP 3 leads to termination of the call; otherwise, call processing enters PIC 3.

PIC 3 — Collect_Information: This is the point in the call where the dialing string is collected from the caller. In addition to DP 21, PIC 3 has two exit means: through DP 4 (Collect_Timeout) to PIC 11, or through DP 5 (Collected_Information) to PIC 4. If the format of the dialed string is incorrect, or the activity is timed out, DP 4 is set and the call is terminated. Otherwise, the call enters the PIC 4 through DP 5.

PIC 4 — Analyze_Information: At this point, the complete dial string is being translated to a routing address. In addition to DP 21, PIC 4 has two exit means: through DP 6 (Invalid_Information) to PIC 11, and through DP 7 (Analyzed_Information) to PIC 5. If the dial string cannot be successfully translated to a routing address, DP 6 is set and the call is terminated. Otherwise, DP 7 is set and the call enters PIC 5.

PIC 5 — Select_Route: For the routing address obtained in PIC 4, the SSP selects one or more physical routes toward that routing address. In addition to DP 21, PIC 5 has two exit means: through DP 8 (Route_Select_Failure) to PIC 11, and through DP 9 (Route_Selected) to PIC 6. If one or more physical routes cannot be selected (possibly due to focused network congestion), DP 8 is set and the call is terminated. Otherwise, DP 9 is set and the call enters PIC 6.

PIC 6 — Authorize_Call_Setup: Certain service features may restrict the type of calls that may originate on a given line or trunk. In addition to DP 21, PIC 6 has two exit means: through PIC 10 (Authorization_Failure) to PIC 11, and through DP 11 (Origination_Authorized) to PIC 7. If, for any reason, the authorization fails, DP 10 is set and the call is terminated. Otherwise, DP 11 is set and the call enters PIC 7.

PIC 7 — Call_Sent: At this point, control over the establishment of the call has been transferred to the T_BCSM object, and the O_BCSM object is waiting for a signal confirming that the call has been presented to the called party, or that the called party cannot be reached for a particular reason (it may be busy or did not answer the phone). In addition to DP 18 and DP 21, PIC 7 has four exits: If DP 13 (O_Called_Party_Busy) is set, the call is terminated. DP 12 (Route_Failure) is set if the network experiences congestion on the chosen route; in such a case, control passes to PIC 5 so that a different route can be selected for the call. When the called party is alerted, the T_BCSM informs the O_BCSM of this by setting DP 14 (O_Term_Seized); in such a case, control passes to PIC 8. If the T_BCSM informs the O_BCSM that the called party has answered the phone (possibly as a result of a race condition when the party called just happens to pick up the phone to make a call), DP 16 (O_Answer) is set and control passes to PIC 9 (skipping PIC 8).

PIC 8 — O_Alerting: At this point, O_BCSM is waiting for the called party to answer. Besides DP 18 and DP 21, this PIC has two exits: DP 15 (O_No_Answer) is set if the T_BCSM informs the O_BCSM that a call was not answered within a time period known to the T_BCSM. The processing of DP 15 terminates further processing of the call. If, on the other hand, the called party answers within the specified period, T_BCSM sends a respective message to the O_BCSM, which results in DP 16 (O_Answer) being set; call processing now moves to PIC 9.

PIC 9 — O_Active: At this point, the call is active, i.e., the parties are communicating with each other. In addition to DP 18 and DP 21, this PIC has two exits: If the called party hangs up the phone, DP 19 (O_Disconnect) is set and PIC 10 is entered (note that if the calling party disconnects, DP 21 will be set). If the network experiences problems while the call is active, DP 17 (O_Connection_Failure) is set and the call is terminated.

PIC 10 — O_Disconnect: Note that this PIC is reached when the called party disconnects the phone; i.e., the T_BCSM detects that the called party disconnected and sends a message to the O_BCSM about this event. In this PIC, the O_BCSM performs the necessary cleanup work (releasing call resources, etc.) and sets DP 20 (O_Disconnect_Complete) to terminate the call. DP 20 is the only exit criterion defined in this PIC.

PIC 11 — O_Exception: Except for PIC 1 and PIC 10, control from other PICs is passed into PIC 11 as a result of exceptional conditions arising during call processing. During normal processing of the call in other PICs, many resources may have been allocated (the SSP may have established a link with the SSP, trunks or lines may have been reserved for the call, etc.). If call processing fails, these resources need to be de-allocated. This PIC performs the needed cleanup to de-allocate any resources that may have been allocated during normal call processing. At the end of processing the exception and cleaning up the resource state, PIC 11 enters PIC 1 without any DP associated with this transition.

3.1.8   Terminating BCSM (T_BCSM)

The terminating-half call model state machine of the BCSM defined in Q.1204 [ITU92b] contains 8 PICs and 14 DPs. Figure 3.7 depicts the T_BCSM state machine. Note that the numbering of both the PICs and DPs continues from that of O_BCSM, instead of starting afresh. Thus, the first PIC of T_BCSM is numbered 12, and the first DP is numbered 22. As was the case with O_BCSM, some PICs have a common exit criterion. Namely, PICs 13 to 17 have a common exit criterion in the form of DP 35 (T_Calling _Party_Disconnect_and_Abandon). This DP is set if the calling party disconnects the call while it is in the middle of being set up at the T_BCSM, or if the called party prematurely abandons further call processing. Processing of DP 35 always results in a transition to PIC 12. PIC 16 and 17 also have an additional exit criterion in the form of DP 32 (T_Mid_Call). This DP serves the same purpose for the called party as DP 18 did for the calling party, i.e., it implements mid-call services. For the T_BCSM, because exit through DP 32 and 35 is common to many PICs, the description below will omit these and discuss other exit criteria in detail.

> **PIC 12 — T_Null**: This is a catchall PIC that absorbs exceptions resulting from processing the call (see transitions leading from DPs 34 and 35 to this PIC in Figure 3.7). At this point, the call does not really exist; however, a message has been received from PIC 7 of the O_BCSM informing the T_BCSM to set up a call. DP 22 (Termination_Attempt) is set and the control passes to PIC 13.
>
> **PIC 13 — Authorize_Termination_Attempt**: This PIC verifies whether the call is to be passed to the terminating party. PIC 2, which is a counterpart in the O_BCSM to this PIC, ascertains whether the caller is authorized to initiate a call; this PIC establishes that the called party is authorized to receive a call, that its line has no restrictions against this type of call, and that the bearer capabilities of the caller and called party match. Besides DP 35, there are two exit criteria from this PIC: DP 23 (Termination_Denied) is set if the called party is not authorized (or has incompatible bearer

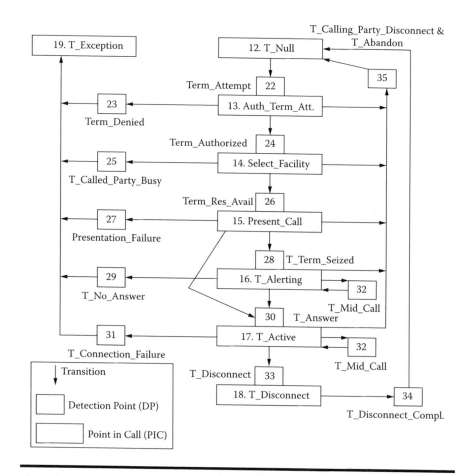

Figure 3.7 Example Terminating Basic Call State Model. (After Figure A.3 in ITU-T, Intelligent Network Distributed Functional Plane Architecture, Recommendation Q.1204, Geneva, Switzerland, March 1993.)

capabilities) to receive the call. In this case, the call is terminated. DP 24 (Termination_Authorized) is set otherwise, and the call proceeds to the next PIC.

PIC 14 — Select_Facility: At this PIC, the terminating resource (i.e., a line or trunk) is selected. Note that the T_BCSM object may reside on an originating or intermediate switch (not only on the terminating one), which means that on these switches it performs the job of finding a trunk to the next switch. Besides DP 35, there are two exit criteria from this PIC: DP 25 (T_Called_Party_Busy) is set if there are no resources to set up a call or if the called party is busy. In this case, the call is terminated. DP 26 (Terminating_Resource_Available) is set otherwise, and the call proceeds to the next PIC.

PIC 15 — Present_Call: At this PIC, the called party is alerted (via AN audible ringing tone) of an incoming call. Besides DP 35, there are three exit criteria from this PIC: through DP 27 (Presentation_Failure), DP 28 (T_Term_Seized), and DP 30 (T_Answer). DP 27 is set if the T_BCSM cannot, for any reason, alert the called party. Processing DP 27 results in the call being terminated. DP 28 is set if the called party is successfully alerted, in which case control enters PIC 16 and the O_BCSM is notified, causing it (the O_BCSM) to go into PIC 8 (O_Alerting). If the called party answers the phone, DP 30 is set and control passes to PIC 17 (bypassing PIC 16).

PIC 16 — T_Alerting: Note that the called party was already alerted in PIC 15; thus, this PIC may seem redundant. However, it serves the purpose of defining an upper bound on how long the called party is alerted. Such an upper bound is needed to prevent indefinite holding of network resources (trunks, lines, or computing resources) that have been acquired to process the call thus far. If within a preconfigured time at the T_BCSM no one picks up the phone, DP 29 (T_No_Answer) is set and the call is terminated. If the called party answers, DP 30 (T_Answer) is set and control passes to PIC 17. DP 35 is set if exceptional conditions warrant the termination of the call. DP 32 is set for mid-call services (control is passed back to PIC 16 on processing DP 32).

PIC 17 — T_Active: At this point, the call enters the active state. As a result of processing DP 30, the O_BCSM is notified so that it also enters PIC 9, its equivalent of the active state. This PIC has four exit criteria. If the network experiences problems maintaining the active call, DP 31 (T_Connection_Failure) is set and the call is terminated. If the called party (i.e., the party corresponding to the T_BCSM) disconnects the call, DP 33 (T_Disconnect) is set and the processing enters the next PIC. If the calling party (i.e., the party associated with O_BCSM) disconnects the call, DP 35 is set and the call is terminated. If a mid-call event occurrs, DP 33 (T_Mid_Call) is set; after the service has been performed, control is passed back to PIC 17.

PIC 18 — T_Disconnect: This PIC is reached when the called party disconnects the call. The T_BCSM sends a message to the O_BCSM, causing the O_BCSM to set DP 19 and enter PIC 10. There is only one exit criterion from this PIC: DP 34 (T_Disconnect_Complete). This DP terminates further processing of the call by releasing any resources accrued thus far in maintaining the call.

PIC 19 — T_Exception: With the exception of PICs 12 and 18, control from other PICs is passed into PIC 19 as a result of an exceptional condition arising during call processing. During normal processing of the call in other PICs, many resources may have been allocated (the SSP may have established a link with the SSP, trunks or lines may have been reserved for the call, etc.). If call processing fails, these resources need to be de-allocated. This PIC performs the needed cleanup to de-allocate any resources that may have been allocated during normal call processing. After the processing of the exception is complete and the resource state has been cleaned up, PIC 19 enters PIC 12 without any DP associated with this transition.

3.2 Service Architecture for the Cellular Public Switched Telephone Network

Until recently, the cellular network expanded with unprecedented growth. In 1984, there were 92,000 U.S. cellular subscribers, compared to approximately 140 million U.S. subscribers on December 31, 2002 [FCC03]. Clearly, the cellular network is an important component of the PSTN.

The current generation of the cellular network is referred to as 2G, short for second-generation cellular network. 2G is a digital voice network; however, its endpoints are not Internet capable. It provides mobility and data transmission in the form of the Short Message Service (SMS).[6] 2G is a precursor of 3G, or third-generation cellular network, that is based completely on IP. However, as was discussed in Section 1.1, much uncertainty surrounds 3G at this time [GAR02, GOO00]. In the meantime, telephone operators and service providers are experimenting with a technology called 2.5G, an *ad hoc* stepping stone between 2G and 3G. 2.5G systems allow cellular handsets to send and receive IP packets in a digital cellular network, in effect turning the cellular handset into an IP device. However, it is important to note that the voice traffic between the 2.5G cellular handset and the network does not utilize the IP connection (i.e., voice traffic is not packetized and transmitted over the IP connection). Instead, certain time slots on some frequencies are reserved for data traffic and the rest of the spectrum is dedicated to transmitting the voice traffic.

Fortunately, the 2G and 2.5G cellular networks are well integrated with the PSTN; at the services layer, 2G and 2.5G are heavily influenced by the concepts of the IN [BER97, FAY97]. Much like the IN, Wireless IN (WIN) is based on an architecture that separates call processing from

---

[6] SMS is a set of services that support the storage and transfer of short text messages (200 bytes or less) through the cellular network.

enhanced feature functionality. Many of the ideas covered in Section 3.1 — the IN conceptual model with its four planes (Section 3.1.3); BCSM, DPs, and PICs (Section 3.1.5); O_BCSM (Section 3.1.7); and T_BCSM (Section 3.1.8) — apply just as well to WIN. To the designers of the IN, mobility was pertinent from the very beginning. CS-1 provided limited support to cellular services; however, CS-2 and beyond integrated the cellular network even more.

There are some differences between the traditional IN and WIN; this section highlights the similarities and differences that are pertinent to this book. A detailed treatment on WIN and the relevant cellular standards can be found in [FAY97, GAL97].

3.2.1 *Physical Entities in WIN*

Figure 3.8 depicts a representative cellular network. As evident, it contains some well-known entities, specifically the SCP, intelligent peripheral, and SN. In addition to these, it has other entities that are pertinent to the cellular network and provide support for executing services in that environment.

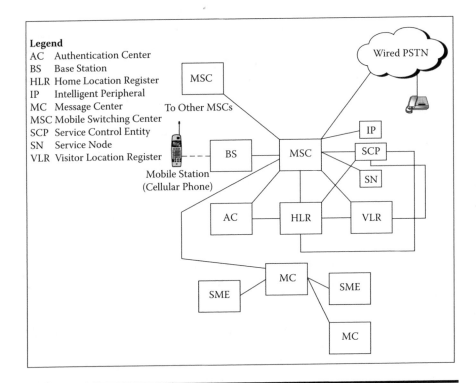

Figure 3.8 Representative cellular network.

3.2.1.1 Mobile Switching Center (MSC)

The MSC is the brain of the cellular network. It is an automatic switching system that shuttles user traffic between the cellular network and the PSTN and other MSCs in the same or different network. It provides the basic switching functions and coordinates the establishment of calls to and from the cellular subscribers (through the base station). The MSC connects the cellular network with the landline (or wired) PSTN. In addition, it also incorporates mobile application functions and other service functions.

3.2.1.2 Base Station (BS)

The BS represents all the functions that terminate radio communications at the network side of a mobile station (also called a cell phone). It controls the radio resources and manages network information required to provide telecommunication services to the mobile station. A BS serves one or more cells (a cell is a geographic area reachable by a signal provided by the BS). The BS incorporates radio functions and radio resource control functions.

3.2.1.3 Authentication Center (AC)

The AC manages and processes the authentication information related to a mobile station. This information consists of encryption and authentication keys as well as complex mathematical algorithms to provide encryption and security when using network services. The AC incorporates database functions used for the authentication keys and authentication algorithm functions. The AC may be located within and be indistinguishable from a home location register. In addition, an AC may serve more than one home location register.

3.2.1.4 Home Location Register (HLR)

The HLR is the primary database repository of subscriber information used to provide control and intelligence in cellular networks. It represents the home database for those who have subscribed to services in that home area. The HLR contains a record for each home subscriber that includes current location information, subscriber status, subscribed features, and directory numbers. Supplementary services or features that are provided to a subscriber are controlled by an HLR. The HLR may be located within and be indistinguishable from an MSC. An HLR may serve more than one MSC.

3.2.1.5 Visitor Location Register (VLR)

The VLR represents the local database's control and processing functions that maintain temporary records associated with individual network subscribers who are away from their home area. A visitor can be a mobile subscriber served by one of many systems in the home service area or a subscriber who is roaming into another service provider's area. The VLR contains the subscriber's current location, status, and service information as derived from the HLR (a transfer takes place between the HLR and the VLR when a mobile subscriber roams into a foreign area). The local MSC consults the VLR to route calls to and from visiting subscribers. The VLR may serve more than one MSC and incorporates database functions, mobile application functions, and other service logic functions.

3.2.1.6 Short Message Entity (SME)

An SME is an entity that can originate short messages for the SMS service, terminate short messages, or both. A mobile station is a good example of an SME; however, even other entities such as HLR, MSC, or an MC could act as an SME.

3.2.1.7 Message Center (MC)

The MC acts as a store-and-forward point for SMS messages. The MC forwards (routes) the SMS messages to the recipient, or if the recipient is unavailable to receive them, the MC can store the SMS messages for the recipient and deliver them when the recipient becomes available.

3.2.2 WIN PICs and DPs

WIN has adopted the call model from CS-2. Because the CS-2 call model is a subset of the Q.1204 BCSM discussed in Section 3.1.7, the DPs and triggers of Q.1204 BCSM are just as applicable to WIN networks as they are to traditional IN networks. There are two differences, however. First, in traditional IN, a static DP cannot be unarmed dynamically; the switch has to be provisioned to do so. This elimination has been restricted in WIN, so DPs in WIN are dynamically armed or disarmed. Second, in WIN, the BCSM alone may not be able to properly invoke all services. Other entities may provide additional pieces of information to the service platform. A good example of this is registration. The BCSM does not contain any states for registration and de-registration of mobile stations. WIN standards define a separate finite-state machine for location registration

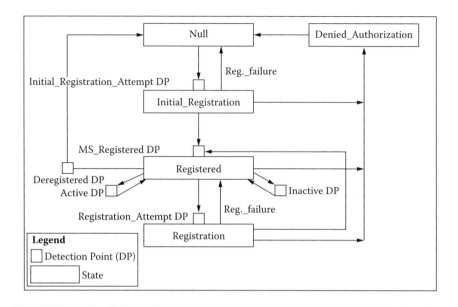

Figure 3.9 WIN location registration function state register.

function. Figure 3.9 reproduces (from [FAY97]) the state machine for the location registration function. This state machine has six DPs that can be used for the IN services.

To summarize, service creation and execution in the PSTN are accomplished by the use of the IN conceptual model. The IN conceptual model defines the entities that participate in the service and their respective roles. It also defines the protocol used by these entities for intercommunication. A salient point to note about the PSTN service architecture is that the protocol used for service execution is different than the one used for signaling session setup. While the session setup is accomplished by a protocol called ISUP, service execution is accomplished by another protocol called Transaction Capabilities Application Part (TCAP) (or Intelligent Network Application Part (INAP)). Both of these protocols are transported over the packet-based SS7 signaling network. Our work abstracts the details of these protocols, so we discuss them in a minimal manner when encountered; interested readers can consult Russell [RUS02] for more information on this topic.

3.3 Service Architecture for Internet Telephony

Although, as observed, the service architecture for the wireline and cellular PSTN is well specified and stable, this is not the case for Internet telephony.

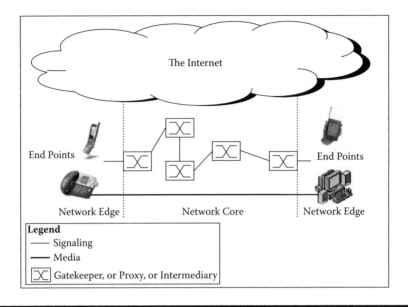

Figure 3.10   Internet telephony architecture.

Service architecture for Internet telephony is still in an embryonic phase [DIA02, GLI00, LEN04a]; much more work needs to be conducted before a stable service architecture is defined. Figure 3.10 depicts the entities that participate in an Internet telephony service; the loose collection of these entities can be characterized as a possible architecture for Internet telephony. Figure 3.10 depicts a pure Internet environment, i.e., no interaction with the PSTN is considered.

Compared to the architecture of the PSTN depicted in Figure 3.1, Figure 3.3, and Figure 3.8, the architecture of Figure 3.10 is extremely simplistic. Unlike the PSTN, where the signaling traffic uses a separate network from the media traffic, in Internet telephony the same physical network is used for both media and signaling. The signaling messages are routed through the core of the Internet by intermediaries often called a gatekeeper or a proxy. Once the signaling messages establish a session successfully, media flows directly between the endpoints, bypassing the intermediaries. This is another departure from the PSTN where media between the two end switches flow between intermediaries, including the tandem switches. Also note in Figure 3.10, unlike the IN in the PSTN, there is no centralized service execution platform for Internet telephony. These differences simply reflect the end-to-end nature of the Internet and the reality that Internet endpoints (personal computers, laptops, personal digital assistants, 3G phones) are far more powerful and capable than PSTN endpoints (simple phones with a 12-digit keypad and a display window).

3.3.1 *Service Specification in Internet Telephony*

Three protocols dominate in Internet telephony. At the signaling layer, Session Initiation Protocol (SIP) and H.323 are used to set up, maintain, and tear down communication sessions as well as provide services. At the media layer, RTP is used to transport the voice samples between the communicating users. As a prerequisite to discussing the service architecture in Internet telephony, it is instructive to study the technologies used to create Internet telephony services (Lennox [LEN04a] provides an indepth treatment of this topic). Services in Internet telephony can be created using SIP Common Gateway Interface (CGI) [LEN01a], the Call Processing Language (CPL) [LEN04], or SIP servlets [JAV03]. Two of these (SIP CGI and SIP servlets) are applicable to SIP only and not H.323. Even though SIP and H.323 are not formally introduced until Chapter 4, it is instructive to note the interdependence between the signaling protocols and the technologies used for service creation in Internet telephony.

3.3.1.1 *SIP CGI*

Patterned heavily after the Hypertext Transfer Protocol (HTTP) CGI, SIP CGI allows a SIP intermediary to execute a CGI script that contains the service to be implemented. The script itself is language independent; it may be written in C, C++, Perl, Tcl/Tk, Python, or even the UNIX shell. SIP CGI standardizes the interfaces between the intermediary and the script. SIP CGI is targeted at experienced and trusted developers. Because the script executes within the same process space as the intermediary, a malfunctioning script can adversely affect the intermediary. Furthermore, SIP CGI is specific to the SIP signaling protocol only. For these reasons, its usefulness as a service creation tool is limited.

3.3.1.2 *CPL*

CPL primarily targets the end users and is applicable to both SIP and H.323. A CPL script describes a service, and once it is constructed, it is uploaded to the endpoint (or an intermediary). When a call setup message arrives at the endpoint (or intermediary), the script is executed. Unlike SIP CGI, CPL is designed to be safe; the language used to describe CPL scripts lacks loops and function calls. Furthermore, it does not allow access to any external programs. Ironically, these attributes are viewed as a drawback because they inhibit the creation of advanced services that require access to external data sources or a more powerful programming paradigm.

3.3.1.3 SIP Servlets

As was the case with SIP CGI, SIP servlets are inspired by equivalent counterparts in HTTP. A SIP servlet is a Java-based component managed by a servlet engine. The servlet engine is part of the SIP intermediary hosting the servlet. Because SIP servlets are written in Java, the security model of the Java sandbox ensures that a malfunctioning servlet does not harm the host itself. The main disadvantage of SIP servlets is that they are tied intimately to the signaling protocol (SIP) used as well as the programming language (Java) in which an Internet telephony entity is written.

3.3.2 Service Residency in Internet Telephony

In the PSTN, services reside at the core of the network because the core, consisting of the IN entities, is much smarter than the edge of the network, which contains simple endpoints. The Internet is characterized by endpoints that are far more intelligent. Unarguably, a personal computer or an Internet-capable phone is far more powerful and expressive than the 12-button phone. The core of the Internet is fairly simplistic in that it performs routing services only. All other services, including data presentation, are done by the powerful endpoints that reside at the edges of the network. Due to the lack of a centralized control point, and compounded by the fact that Internet endpoints can host and run services themselves, problems already prevalent in traditional PSTN, such as feature interaction, become more pronounced in the Internet. Feature interaction results when one feature modifies or influences other existing features in defining the overall system behavior; some feature interactions are conducive to the overall service experience, but many are not. This type of interaction is explored in more detail in [LEN04a, LEN00].

The characterization of intelligence at the edges implies that the best place to deploy Internet telephony services is at the edges, and, indeed, this has been done with mixed results [LEN04a]. But not all services are best executed in an endpoint model. Services such as starting a conference call when all parties in a call are deemed to be present are best done by a centralized entity who knows of the state of the participants in the call. The optimum place to deploy Internet telephony services is still a topic of ongoing research [GUR03c, VAN99, WUX00, WUX03].

Chapter 4

Comparative Analysis
of Signaling Protocols

As is the case with any distributed system, a protocol is required to synchronize the attendant entities for deterministic behavior. We list the properties that are desirable in such a protocol and analyze three signaling protocols — Bearer Independent Call Control (BICC), H.323, and Session Initiation Protocol (SIP) — to choose one that can serve as a candidate protocol for our work.

4.1 Desirable Properties of a Candidate Protocol

Perhaps the most critical function of a signaling protocol is to enable services beyond normal call establishment. All telephony signaling protocols contain primitives for call establishment; however, the richness of a protocol can be gauged by the support of primitives that enable other services (call waiting, call transfer, hold, floor control, unified messaging, interactive voice response, etc.). The primary value proposition and an important benefit of Internet telephony is the ability to deliver a wide range of new services, especially working in conjunction with traditional telephony and other communication technologies, such as the Web, instant messaging, and presence. To this extent, we list the desirable properties of a candidate protocol.

4.1.1 Widespread Acceptance

Our work detailed in subsequent chapters depends on the extensive availability of Internet telephony endpoints. Such endpoints may support many signaling protocols (although not simultaneously). Thus, the first critical property is a high acceptance rate marked by the widespread deployment of the Internet telephony endpoints that use the candidate protocol.

4.1.2 Protocol Expressiveness

We define protocol expressiveness as the degree of support provided by the protocol for expressing the capabilities about other services besides session setup, which is the most primitive requirement of any signaling protocol. In our work, services such as presence, instant messaging, and asynchronous event notifications play an important part. A candidate protocol must not only support setting up voice sessions, but also be expressive enough to enable other types of services.

4.1.3 Protocol Extensibility

The third property we desire in a candidate protocol is extensibility; specifically, the protocol should be elastic enough to support extensibility in two ways. First, it should be possible to *transport* arbitrary descriptive elements in signaling (we will use this in Chapter 6), and second, it should be possible to *describe* what is being transported.

4.1.4 Primitives for Capability Description and Negotiation

Closely tied to extensibility is the need for capability description and negotiation. Our work in Chapters 6 and 7 requires that the communicating entities be capable of describing their individual capabilities to each other. We are not as much concerned with the description and negotiation of media as with the description and negotiation of the capabilities of the *endpoint* itself. For example, a sender might wish to describe for the receiver the payload being transported as well as the type of signaling messages that it supports. The candidate protocol must be flexible enough to accommodate these needs.

4.1.5 Transaction-Style Message Exchanges

Another property of a candidate protocol is a simple transactional, request–response-driven signaling that has proved durable on the Internet (witness the success of Hypertext Transfer Protocol (HTTP), File Transfer Protocol (FTP), etc.). A request–response property in the candidate protocol will also aid in synchronizing the entities on the public switched telephone network (PSTN) and Internet Protocol (IP) networks.

4.1.6 Support for an Event-Based Communications Model

The candidate protocol must also support an event-based communication model. Such a communication model is far more amenable to large-scale systems. Our work in Chapter 6 depends on key events in the PSTN migrating to the Internet for service execution. Such a system cannot be constructed based on a synchronous communication model because it imposes a tight coupling between the involved participants.

4.1.7 Support for a Flexible Naming Scheme

The final property of a candidate protocol is support of a flexible naming scheme. Resources in the PSTN are identified by numbers, but in the IP network, resources can be identified using a much richer vocabulary, which includes names, numbers, domains, etc.

4.2 Protocols Evaluated

We evaluated three protocols: International Telecommunication Union — Telecommunications Standardization Sector's (ITU-T) BICC, ITU-T's H.323, and Internet Engineering Task Force's (IETF) SIP.

4.2.1 BICC

ITU-T's BICC [ITU00a] is a signaling protocol based on narrowband Integrated Services Digital Network (ISDN) User Part (ISUP); it is used to support narrowband ISDN services over a broadband backbone network without interfering with interfaces to the existing network or deployed services. BICC's main purpose is to disassociate the bearer — which may be an Asynchronous Transfer Mode (ATM) network, IP network using the Real-Time Transport Protocol (RTP), or the existing time division multiplex (TDM) network — from the signaling. BICC signaling can be used to establish a voice session over any bearer technology, including the current TDM network.

BICC has its roots in ISUP, and it attempts to correct the deficiencies of ISUP. ISUP messages carry both call control and bearer control information, identifying the physical bearer circuit used to set up a call. However, a circuit is specific to the current PSTN TDM network; thus, ISUP is closely tied to the bearer network used. It cannot be used to set up sessions across other networks, such as ATM or IP. ITU-T invented BICC to correct this shortcoming.

Because of the fact that BICC is closely tied to ISUP-based networks, it cannot deliver services beyond existing PSTN-type features. Even when BICC is used for an IP network, the primitives in the protocol are not expressive enough to render its use for other services beyond those already present in the PSTN. BICC's strengths are its ease of interoperability with PSTN and the transparent delivery of existing circuit-switched services to Internet endpoints.

4.2.2 H.323

H.323 [ITU03] was the first ever widely used Internet telephony signaling protocol. Reflecting on its roots in the PSTN domain rather than the Internet, H.323 maintains implicit preference of the latter network; for instance, its call model is a derivative of the Q.1204 call model we described in Chapter 2. H.323 was first standardized by the ITU-T in 1996, with revisions following in 1998, 2000, and 2003. Currently, H.323 is the most widely deployed Internet telephony protocol.

H.323 is an umbrella protocol; it contains a set of individual protocols that perform the task of call signaling, media negotiation, speech encoding, data transport, and feature specification. Some of these protocols are taken from existing ITU-T protocols, while the remaining are made up from the IETF standards. Figure 4.1 depicts the positioning of the protocols in an H.323 stack.

H.323 uses IP; thus, all the individual protocols employ IP for transport and delivery. On the media side, H.323 standardizes codecs for audio and visual components. A variety of audio codecs are provided, ranging from the telephone speech quality G.711 codec (8000 samples/second with an 8-bit sample to yield an uncompressed speech stream at a rate of 64 Kbps) to the highly compressed G.723.1, which produces a speech stream at a rate as low as 5.3 kbps. Video codecs include the low-bitrate encoding H.263 and high-compression H.264 standards. The codecs digitize H.323 media (audio and video) to be transported using the RTP and RTCP.

Because multiple compression algorithms are supported, H.323 provides a protocol called H.245 for endpoints to negotiate the codecs. ITU-T's Q.931 call control protocol is used for establishing and releasing connections, providing dial tones, indicating call progress to the endpoints, and providing for other standard telephony-related messages. H.323 networks consist of endpoints (called terminals) and intermediaries (called

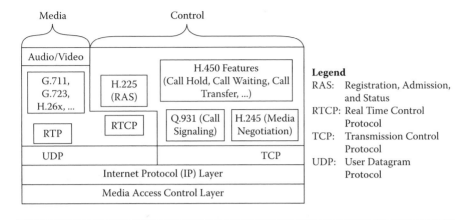

Figure 4.1 The H.323 protocol stack.

gatekeepers). A terminal authenticates itself and gets the zone it belongs to from the gatekeeper. (A collection of terminals and gatekeepers is called a zone; many such interconnected zones make up a wider area H.323 network, and many such networks provide a global H.323 network.) The gatekeeper also controls other resources, such as bandwidth requested by the terminal, registering the address of the terminal so it can be reached by other users, etc. All these workings are captured in the H.225 (Registration/Admission/Status) protocol.

Finally, features or services in H.323 are specified by the ITU-T H.450 set of standards. Each feature has an associated protocol defining it; for instance, H.450.2 specifies call transfer and H.450.4 specifies call hold.

4.2.3 SIP

The Session Initiation Protocol is an application-layer protocol used to establish, maintain, and tear down multimedia sessions. It is a text-based protocol with a request–response paradigm modeled after other successful Internet protocols, like HTTP, FTP, and Simple Mail Transport Protocol (SMTP). Figure 4.2 depicts the overall SIP suite.

At the media layer, SIP is indistinguishable from H.323; both use RTP, RTCP, and the G.7xx codec schemes. At the signaling layer, SIP is fashioned after other successful Internet protocols like HTTP and SMTP, which are all text based. A SIP entity, like a HTTP browser, issues requests to a server, which returns responses. A sequence of such requests and responses is called a transaction in SIP. A SIP ecosystem consists of network intermediaries (such as proxy servers, registrars, and redirect servers) and endpoints. SIP endpoints are called user agents.

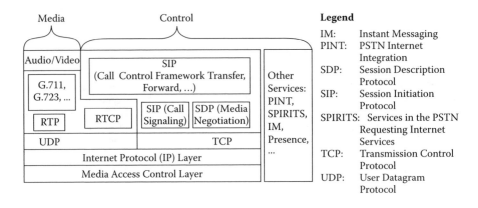

Figure 4.2   The SIP stack.

There are two types of SIP user agents: a user agent client (UAC) and a user agent server (UAS). A UAC and a UAS are software programs that execute on a computer, an Internet phone, or a personal digital assistant (PDA). They are utilized by the *physical* user in possession of that computer, Internet phone, or PDA to initiate and receive a phone call. A UAC originates a request (i.e., starts a phone call) and a UAS accepts and acts upon a request. User agent servers typically register themselves with a registrar, which binds their current IP address to an e-mail-like identifier used to identify the user (e.g., the identifier vkg@iit.edu can be bound to a certain IP address during registration). This registration information is used by SIP proxy servers to route the request to an appropriate UAS.

Proxy servers are SIP intermediaries that provide critical services such as routing, authentication, and forking. Forking is the ability of a SIP proxy to branch an incoming request into multiple outgoing requests, each targeted to a different UAS. A complex side effect of forking is receiving and making sense of the many responses arriving from each of the downstream user agent servers, and sending one response to the upstream UAC.

A SIP proxy, upon the receipt of an incoming call setup request, will determine how to best route the request to a downstream UAS. If the request corresponds to a user present in the domain that the proxy is authoritatively responsible for, the proxy will consult the location service to determine the user's location. The location service is updated by the registrar when a UAS (under the control of the user) registers with the registrar. If the user is indeed registered, the proxy will transmit the request downstream toward the UAS. If the user agent is not registered (or some other error occurs), the proxy will reject the request by issuing a response. A SIP proxy also handles requests going to other domains. If the proxy determines that a call is being set up with a user that is not in the domain that the proxy is responsible for, it will consult the Internet Domain Name Service (DNS) to route the request to its destination domain. When the request reaches the destination domain, the proxy responsible for that domain will further act upon it by consulting the location server and forwarding the request appropriately.

By far, the most critical service provided by proxies is routing; a proxy routes a request from the source to the destination using a variety of techniques, ranging from simple DNS lookups to executing Call Processing Language (CPL) documents, Common Gateway Interface (CGI) scripts, and SIP servlets [GUR01].

The request to establish a session in SIP is called an INVITE. An INVITE request generates one or more responses. Responses to requests indicate success or failure, distinguished by a status code. Responses with status code 1xx (100 to 199) are termed provisional responses and serve to update the progress of the call; the 2xx code is for success and higher numbers are for failures. Responses with codes 2xx to 6xx are termed final and serve to complete the INVITE request. The INVITE request is forwarded by a proxy (through possibly another chain of proxies) until it gets to its destination. The destination sends one or more provisional responses followed by exactly one final response. The responses traverse, in reverse order, the same proxy chain that the request did. Figure 4.3 provides a timeline diagram of call establishment and teardown between a UAC and a UAS. The request is forwarded through a chain of two proxies.

With reference to Figure 4.3, the UAC sends an INVITE to P1. It is now the responsibility of P1 to route the call further downstream, as discussed

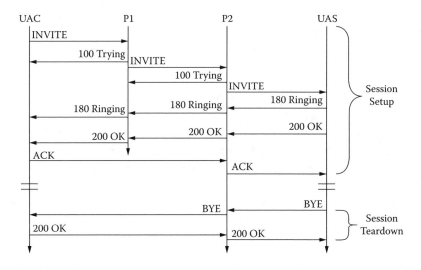

Figure 4.3   SIP call establishment and teardown.

above. From the UAC's reference, P1 is called an outbound proxy. P1 determined that the request should be forwarded to P2 (the UAS is in a different domain). When the request arrived at P2, it queries its location server and further proxies the request to the UAS. From the UAS's point of view, P2 is the inbound proxy. The UAS issues a provisional response followed by a final response. The call is set up when the UAC receives the final response (200 OK) and sends out the ACK request. Note that the ACK request (as does the BYE request) travels through P2, but not P1. This is because SIP allows intermediaries to only participate in a session as much (or as little) as they would like to. Once a session is set up, P1 is no longer interested in being part of subsequent signaling, whereas P2 is. Thus, subsequent requests beyond the INVITE always traverse P2. If P2 had not indicated an interest in being part of subsequent requests, signaling would have occurred directly between UAC and UAS after the session setup.

Besides the proxy server as a network intermediary, SIP also has another entity that has been used as a powerful network intermediary — a Back-to-Back UA (B2BUA). From a high-level point of view, a B2BUA is comprised of two SIP user agents connected together. One half of the B2BUA receives a SIP message, translates the semantics of the SIP message internally, and then initiates a new outgoing SIP message from the other half. At the basic level, the B2BUA functions semantically like a SIP proxy, passing the intent of the SIP messages it receives from one end to the other, but without the restrictions of a proxy on modifying the messages. In addition, as a SIP UA, the B2BUA may initiate SIP messages and perform call control and call management functions. B2BUAs typically run as part of (or, indeed, may comprise) an application server. Depending on the exact nature of the service they are providing, they may perform different functions. For instance, a B2BUA providing third-party call control [CHI02] is distinctly different from one that provides an anonymization service by mangling the "To" and "From" headers.

In addition to the INVITE request that sets up a session, SIP includes requests to tear down a session (BYE) and register an endpoint with the network (REGISTER). The protocol is extensible and has been extended to support services such as transporting instant messages [CAM02a] and providing a framework by which SIP nodes can asynchronously request notification from remote nodes indicating that certain events have occurred [ROA02].

A simple SIP request is depicted in Figure 4.4. A SIP request, as well as a SIP response, is composed of two discrete parts: a list of headers and an optional body. The body is delimited from the headers by an empty line.

The headers describe various capabilities of SIP, such as the types of methods that the user agent supports (Allow), the sender of the request

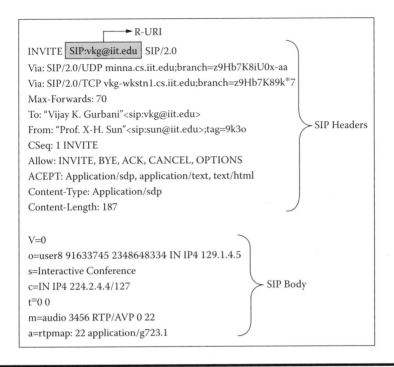

Figure 4.4 A SIP request.

(From), the recipient of the request (To), the number of intermediaries that have handled the request (the "Via" list), the type of Multipurpose Internet Mail Extension (MIME) types it can accept (Accept), and the MIME type of the SIP body encoded in the SIP message (Content-Type). MIME [FRE96a] is an Internet standard that uses well-known and globally unique tokens called MIME types to describe bodies exchanged in an Internet protocol such as SIP or SMTP. A special header called "Content/Type" contains an Internet Assigned Numbers Authority (IANA)[1]-registered MIME type that describes the body. For example, "Content/Type: application/sdp" describes a SIP message that transports Service Data Point (SDP) information in the body (SDP is an IETF standard [HAN98] that describes the media capabilities of a user agent, including the codecs supported and the IP addresses and port numbers where the media are destined to). MIME defines mechanisms for sending arbitrary types of information

---

[1] IANA (http://www.iana.org) is the organization that is well known for having overseen the allocation of IP addresses to Internet service providers. In addition to that, IANA also has the responsibility to maintain all unique parameters and protocol values required for the operation of the Internet. These include port numbers, character sets, and, of course, MIME values.

objects in an application-layer protocol such as SIP, HTTP, or SMTP. The exact object a protocol is carrying is denoted by the "Content-Type" header field. Because an information object may contain binary data, the MIME standard also defines a set of methods for representing binary data in a textual format.

The topmost line of a SIP request contains a special sequence of characters called a Request-Uniform Resource Identifier (R-URI), which represents the ultimate destination of the request. SIP routing is performed by analyzing the R-URI to determine if the resource specified therein matches the resource the entity processing the request is responsible for. If that is the case, the request is consumed by the entity and a final response is issued. Otherwise, if the entity processing the request is a proxy, the request is retargeted and routed further downstream.

Figure 4.5 contains a SIP response. The topmost line of a SIP response is composed of a number ranging from 100 to 699. The 100-class responses are called provisional responses in SIP; 200- to 699-class responses are called final responses. Final responses end a request. A request may elicit many provisional responses, but exactly one final response. The 200-class responses are called successful final responses, 300-class responses serve as redirectors (i.e., they redirect the UAC to try alternate locations), 400-class responses indicate a malformed request, 500-class responses indicate

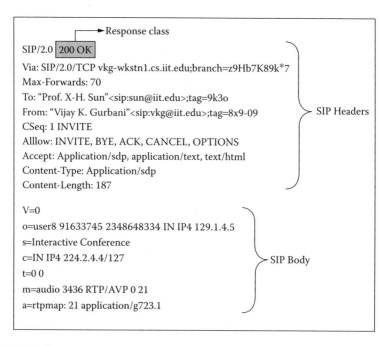

Figure 4.5   A SIP response.

that an error in processing occurred at the entity that received the request, and 600-class responses indicate a global failure in finding the resource indicated by the R-URI.

As Figure 4.5 depicts, a response contains many of the same header fields that the request did. For the INVITE request, a successful response may include a body containing the SDP of the sender of the response; however, the protocol does not mandate that all responses include a body. Whether to send a body depends on the particular semantics defined for a specific request.

4.3 Comparative Analysis

Table 4.1 contains a matrix depicting the comparative strengths and weaknesses of the protocols we evaluated. The rows of the matrix correspond to the desired properties of a candidate protocol we presented in Section 4.1, and the columns contain the protocols we evaluated.

Table 4.1 Comparative Analysis of Evaluated Protocols

	Protocol		
Desired Properties	*BICC*	*H.323*	*SIP*
Widespread Acceptance	No	Yes	Yes
Protocol Expressiveness	No	Limited	Yes
Protocol Extensibility	No	No	Yes
Capability Description and Negotiation	Limited	Yes	Yes
Transaction-Style Message Exchanges	Yes	Yes	Yes
Flexible Naming Scheme	No	No	Yes
Event-Based Communications	No	No	Yes

H.323 and SIP are more widespread than BICC. Early Internet telephony endpoints readily supported H.323, and although the protocol is still used, SIP is fast becoming the preferred protocol for reasons we discuss later. BICC, in all fairness, is not a protocol that an Internet telephony endpoint would run natively; instead, it is used in the core of the network by the telephone switches.

Earlier, we defined protocol expressiveness as how amenable the protocol is in supporting other services besides session setup. For the work discussed in Chapter 5, our requirements of protocol expressiveness include the capability to readily map information elements between the circuit-switched network elements and the Internet telephony endpoint. For the work discussed in Chapters 6 and 7, the protocol must enable asynchronous event notifications from one network to another and, in addition, support Internet-style services such as instant messaging and presence. Clearly, BICC falls short; because it does not run on Internet endpoints, its utility to our work is somewhat limited. H.323 can support services in addition to normal call setup; these services are called supplementary services in H.323. However, such supplementary services — enumerated in H.450 — are geared primarily toward traditional telephony. H.323 is applicable to portions of our work; however, we note that it does not natively support asynchronous event notification, and it does not support Internet-style services such as presence and instant messaging. SIP, by contrast, has provisions to support asynchronous event notifications [ROA02], presence [ROS04], and instant messaging [CAR01].

BICC and H.323 have their roots in telephony and are thus not extensible beyond that domain. Their call models are primarily geared toward establishing, maintaining, and terminating telephone calls. Although SIP can be, and has been, used for the same purpose, its design transcends its use as a Session Initiation Protocol only. SIP can transport arbitrary payload in its signaling messages to allow the establishment of gaming sessions, video sessions, and text chat sessions. It fosters extensibility in two ways:

1. By using new MIME types to describe the payload transported
2. By making it relatively easy to define new behavior in terms of new SIP requests and response codes that the protocol designers did not envision, and, in fact, could not have envisioned, when the protocol was created

All three protocols support capability description and negotiation; however, BICC and H.323 only allow media-related capabilities to be described and negotiated (i.e., codecs supported). They do not allow the endpoint to describe the features that it supports — features that may indicate what services the endpoint can provide. SIP, by contrast, allows all types of attributes to be described and negotiated. Through its use of SDP transported as a payload, it allows the description and (a limited) negotiation of media attributes, and through signaling ("Allow," "Supported," "Require," and "Proxy-Require" headers), it permits the endpoint to describe the set of features it supports, and thereby the services it is capable of providing.

All three protocols allow for transaction-style message exchanges. BICC and H.323, following their genealogies, use a transaction style reminiscent of PSTN signaling. SIP, keeping true to its genealogy, uses a transaction style that closely resembles HTTP, SMTP, and other Internet protocols.

Resources in the PSTN are identified by numbers, but in the Internet, resources are identified using a much richer vocabulary, which includes numbers, names, domains, etc. BICC does not provide a naming scheme beyond the one used in current PSTN to address resources on the network. Endpoints are identified by a collection of numbers, which are interpreted for routing a call by the telephone network in a particular country. Compared to BICC, H.323 does have a much more flexible naming scheme. Resources in an H.323 network can be named by an e-mail-like H.323 URI [BER98][2] or a string of numbers representing a PSTN endpoint. In addition, H.323 also has the concept of an alias, which is simply an easy-to-remember sequence of alphanumeric characters representing an H.323 URI or a phone string. In contrast to H.323 and BICC, SIP has the most flexible naming scheme, which subsumes the naming schemes of the other two protocols. In SIP, a resource can be identified by an e-mail-like URI, an H.323 URI, or a tel URI [SCH04b], the last of which corresponds to a resource assumed to be on the PSTN. The SIP specification contains instructions on how to convert the identifiers using different naming schemes into a SIP URI.

4.4 The Novel SIP-Based Approach

Based on the desired properties of a target protocol and the comparative analysis among the candidate protocols outlined in the previous section, we selected SIP as our protocol of choice. In a sense, Internet telephony protocols like SIP provide a richer palette to work from in our problem domain because they are already better tuned toward multimedia communications. As a signaling protocol, SIP is expressive enough to readily map information elements between the circuit-switched network elements and the Internet telephony endpoints. SIP also possesses built-in support for asynchronous event notification and enables services like presence and instant messaging, which we view as vital components of crossover services. We will cover SIP's support for asynchronous events in more detail in Chapter 6.

To be balanced, SIP has its disadvantages. It is a text-based protocol with an expansive context-sensitive grammar that allows wide latitude in representing headers. This makes it challenging to construct fast SIP parsers [COR04]. Binary protocols like H.323 and BICC leave little room for ambiguity while encoding and decoding the information elements of the protocol, thus yielding a fast-parse cycle. There are also a number of open

---

[2] A URI is a compact string of characters used to name an abstract or physical resource on the Internet; sip:vkg@iit.edu and http://www.iit.edu are both URIs.

challenges in SIP [GUR04b] that are being worked within the standards bodies. As the protocol and implementations mature, these issues will be addressed. In the final analysis, for our work, the benefits of the protocol far outweigh the disadvantages.

Chapter 5

Crossover Services Originating on the Internet

This chapter discusses the first of two crossover services; the events for this type of service originate in the Internet, but the service itself resides in and is executed on the public switched telephone network (PSTN).

5.1 Introduction

The intrinsic value of a computer network is measured by the services it provides to its users. As the number of such networks increases, so do the chances that services residing on one network will need to be accessed by users on a different network. As discussed in previous chapters, the two networks that make up the communication network — PSTN and the Internet — are converging, which necessitates access to services residing in one of these networks from the other.

This cross-service access poses a number of problems to be solved. To formulate the problem set consistently, we will define some terms first. We call the network on which the service runs natively the *local* network (or local domain); alternatively, a *foreign* network (or foreign domain) is one from which a request to execute the service is made. The service and its associated data reside on the local network.

The first problem in accessing services from a foreign domain is that of differing network protocols. The PSTN and Internet use dissimilar network protocols and, in fact, are designed with different goals. While the PSTN is a highly tuned network to transport voice, the Internet is a generalized network that can transport any type of payload — voice, video, or data (text). Second, a request for service arriving from a foreign domain will start service execution in the local domain; as such, the entities in the local and foreign domains need to be synchronized. Third, when a service is accessed from a foreign domain, the semantics of the service must be preserved. (That is, the foreign network may have more capabilities — or fewer — than the local network; in either case, the service should function at a minimal acceptable level.) Finally, local networks that host services have already addressed important issues such as scalability and reliability. There is a temptation to *port* services to the foreign network and then revisit the same issues in context of the foreign network. Instead, we believe that the services and their associated data and procedures are best left in the local network, with some technique created to allow access to these services in a transparent manner from the foreign network.

In this chapter, we describe a technique to address the problems outlined above. The technique is most applicable to domains where entities requesting and executing a service follow a finite-state machine of some sorts. Call controllers — entities that are responsible for setting up, maintaining, and tearing down a voice call, or a multimedia session — in PSTN and Internet telephony readily subscribe to finite-state machines. Thus, the telecommunications domain provides us a rich palette on which to focus the work described in this chapter.

5.2 Motivation

To appreciate the need to access existing PSTN services from Internet endpoints, consider that the majority of services that end users are accustomed to — call waiting, 800-number translation, caller ID, etc. — reside on the PSTN. Users on an Internet telephony endpoint should be able to avail themselves of these services in the same transparent manner that they do when using a traditional PSTN handset. There are three ways to accomplish this for the Internet telephony user, as we discuss next.

5.2.1 Rewrite Services for Internet Telephony

The easiest, albeit the most intrusive, manner to make PSTN services available in Internet telephony is to rewrite all the existing PSTN services for the Internet environment. Although technically feasible, this is not a good solution. It takes anywhere from 6 months to a year to get a PSTN service specified, implemented, tested, and deployed. This already assumes a stable service delivery infrastructure as it exists in the PSTN. Internet telephony, being a new medium, does not as yet have a well-specified service architecture that can be leveraged to deploy new services. The service architecture for Internet telephony is in the early stages of being proposed, as we discovered in Chapter 2. All these factors make it extremely difficult and, in fact, undesirable, to replicate existing PSTN services from the ground up in the Internet domain.

5.2.2 Using a Platform-Neutral Service Creation and Execution Environment

In the PSTN, service logic programs are often written in a specific, often proprietary, language and are designed to run on a specific execution platform. Services do not work across different hardware platforms, even if all the platforms are owned by the same vendor, underscoring the difficulty a customer would face in making a service work across vendor boundaries.

Instead of writing the same service for every network processor, in 1992 AT&T Bell Laboratories assigned researchers to study whether a language-neutral service creation and execution framework would be a feasible option. The research proved that it could indeed be done and culminated in a proposal for a new language, Application-Oriented Parsing Language (AOPL) [SLU94]. AOPL specified a grammar and methodology that provided the service creators with platform-neutral building blocks to create services. Services were to be written in a platform-neutral language and would be compiled into the native language of the platform where the service was to be executed. The service logic would first be compiled to produce a parsing tree, which would then be run through a code generator for a specific machine to produce the binary file that constituted the service. Given a standardized representation of a parsing tree proposed by AOPL, code generators for various architectures could easily be written [FAY97]. Although AOPL proved that this could indeed be done, industry interest in AOPL was simply not there to push it toward a standard; thus, efforts in standardizing it waned [LUH01] as time progressed.

5.2.3 Exploring New Techniques to Reuse Existing Services

A final option for accessing PSTN services in a transparent manner is to devise a technique such that services running on a local domain can be accessed transparently from foreign domains. This preserves the (tested and) deployed service infrastructure in the local domain, while at the same time allowing transparent and scalable access to the service from the foreign domain. Service porting or rewriting is not necessary, as the service can be accessed in a network-agnostic manner. In the next section, we present such a technique, which we term call model mapping with state sharing (CMM/SS).

5.3 Call Model Mapping with State Sharing (CMM/SS)

The technique of CMM/SS depends on and assumes the availability of a call model. Recall from Figure 2.4 that a call model is a deterministic finite-state machine (FSM). States in the FSM represent how far the call has progressed at any point in time. The current state plus a set of input stimuli transition the FSM to the next state. In telecommunication signaling, these input stimuli consist of timers firing and the arrival or departure of signaling messages resulting in the execution of significant events. Events cause transition into and out of a particular state.

As discussed in Section 3.3, call models are already an intrinsic part of telecommunication signaling protocols. The PSTN/Intelligent Network (IN) call model consists of 19 states and 35 input stimuli, and the Internet telephony signaling protocol (SIP) consists of 8 states and 20 input stimuli (Figures 5 and 7 of reference [ROS02]). Besides providing a uniform view of the call to all involved entities, call models also serve to synchronize these entities.

CMM/SS consists of mapping the call model of a foreign domain to that of a local domain so that the foreign domain can access services resident in the local domain. Call mapping is fairly prevalent in the telecommunications domain [CAM02, SIN00, VEM02]; however, it has been used so far simply as an interworking function for signaling and setting up a media path between different networks (i.e., between PSTN and Internet, between SIP and H.323, etc.). Our work extends this use of call mapping in two ways: First, the thrust is not on simply making a call across different networks, but *accessing services* across different networks. Second, consider that service execution in a heterogeneous network implies saving the state of the call in the foreign domain until the service has executed in the local domain. Thus, the state of the call needs to be shared between the foreign domain and the service execution function of the local domain. The CMM/SS technique allows us to do this in a transparent manner. The details of the technique are described next.

5.3.1 CMM/SS: Preliminaries

We begin by formally defining our state space and the call model mapping technique. We then take a look at how the state is effectively shared between the domains to access services.

A *localized state* is a finite tuple s^l of atomic states $s(j)$, with $m \geq 1$, as shown:

$$s^l = s(j)_{j=1}^m = (s_1, s_2, \ldots, s_{m-1}, s_m) \tag{5.1}$$

Let F represent the foreign domain and L the local domain. Both F and L have their own localized states denoted by $F[s^l]$ and $L[s^l]$, respectively.

A *global state* is a finite tuple s^G of localized states $F[s^l]$ and $L[s^l]$:

$$s^G = (F[s^l], L[s^l]) \tag{5.2}$$

Elements of $F[s^l]$ and $L[s^l]$ contain their respective atomic states $s(j)$ from Equation 5.1.

In CMM/SS, the states in $F[s^l]$ need to be mapped to the states in $L[s^l]$. We express the mapping process using the following notation:

$$\left[F \xrightarrow{F[s^l]} L \right] \tag{5.3}$$

i.e., $\forall$ states in F; there exists a mapping between each state in F to an appropriate state in L.

If the notation in Equation 5.3 is expressed as a function $\varnothing(x)$, with x being the domain (or the set of argument values for which $\varnothing$ is defined) of $\varnothing$ and $x \in F[s^l]$, then

$$\varnothing(x) = \begin{cases} y,\ y \in L[s^l] & \text{if two random call models } F \text{ and } L \text{ can be mapped} \\ 0 & \text{otherwise} \end{cases} \tag{5.4}$$

If any two random call models F and L can be mapped to each other, then

$$\varnothing(x) = y,\ y \in L[s^l] \tag{5.5}$$

i.e., $L[s^l]$ is the co-domain of $\varnothing(x)$, which implies that for every state in F, there must exist a possible (maybe nonunique) mapping in L.

5.3.2 CMM/SS: The Technique and Algorithms

In most mappings, the number of states will vary considerably between F and L. It is highly unlikely that two call models with a similar number of states and transitions yield different outcomes in the same domain (telephony, in our case). The task then is to produce a mapping of Equation 5.3 and map F to L.

To map F to L, we start with a state in both call models that has equivalent semantics in both domains. If such a state does not exist, a null state can be introduced that will potentially map to any state as a starting point. Visually speaking, it is as if we have taken two strings of Christmas lightbulbs, each with a varying number of bulbs on them, but both strings having a yellow bulb at the very top. The first stage of call model mapping is aligning the strings such that the yellow bulbs are adjacent to each other. The problem now is how to line up the rest of the bulbs. Before we delve into this, let us first take a look at the reason behind the mapping.

The reason why the mapping is performed is to access services in L from endpoints in F. At various points in the call model states, there will be a need to perform a process we term *service-state handoff*; namely, either F or L, having finished processing the service request as much as it can, hands off the control of the service to the other domain. The receiving domain then continues to process the service until the end of its call model is reached, or another service-state handoff occurs. The states at which a service-state handoff occurs are called *pivot* states. Identifying the pivot states is critical for a successful mapping. The first pivot state will always occur in F because it is the domain that processes the initial session setup message. If processing continues in L until the last state of L is reached, the last state becomes a pivot state for L and a service-state handoff occurs. Once a service-state handoff happens, control in the receiving domain continues from the pivot state that caused the state handoff to the other domain in the first place.

Clearly, F and L will have a differing number of states and transitions between them. Thus, it is not realistic to assume that there will be a one-to-one mapping of the states between F and L, however desirable that may be. Figure 5.1 depicts a mapping of states between F and L, including the two discrete points where service-state handoff occurs: once from F to L and then again from L to F, creating three pivot states.

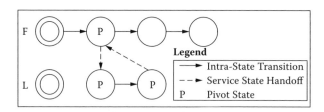

Figure 5.1   Sample mapping.

```
States[] ← {fs₁, fs₂, …, fsₙ₋₁, fsₙ};
Pivot[] ← {fs₁, fs₃ fs₄};

F_CMM_SS(sip_msg)
    for each element e in States[] do
        process sip_msg for state e;
        if (e ∈ Pivot[]) then
                info ← gather information for service state handoff
                info ← service_state_handoff(info);  // May be
                //asynchronous; synchronous behavior shown here
                analyse info;
                transition_to_next_state(e, info);
        else // e is not a member of Pivot[]
                transition_to_next_state(e, null);
        end if;
    done;
```

Figure 5.2   The CMM/SS algorithm for the *F* domain.

CMM/SS includes algorithms to perform service-state handoffs in *F* and *L* domains. Figure 5.2 contains a high-level overview of the algorithm from the viewpoint of the *F* domain. Currently, determining the pivot states in *F* is a manual process and involves studying the call models to settle on which ones will serve as good pivot states. The first pivot state always occurs in *F*, once all the relevant information for the session setup has been obtained. Enough information must be passed from *F* to *L* when a service-state handoff occurs to allow *L* to further process the session.

Figure 5.3 contains an algorithm from the viewpoint of the *L* domain, the domain responsible for providing the actual service. When the first service-state handoff occurs, *F* imparts enough information to *L* to allow *L* to provide services. *L* loads the user's profile and starts providing services (in our case, because *L* is the PSTN/IN, it will initiate the IN call model and arm the detection points (DPs) for service execution). As was the case with determining pivot states in *F*, pivot states for *L* are determined manually by studying the pertinent call model to mine the candidate set.

5.3.3   CMM/SS: State Sharing and Global State

In CMM/SS, state is actually distributed and shared between *F* and *L*. When a call request arrives at *F*, a state transition occurs to state $p \in F[s']$, and processing is temporarily suspended at *F* while a service-state handoff occurs to *L* for service execution (horizontal line from "a" to "1" in Figure 5.4). This initial handoff is a simple mapping of $p \in F[s']$ into an equivalent state $p' \in L[s']$.

```
States[] ← {ls₁, ls₂, …, ls_{n-1}, ls_n};
Pivot[] ← {ls₄, ls₆, ls₈};
curr_state ← null;

L_CMM_SS(service_state_info)
    if (first service state transfer)
        get user information from service_state_info;
        load user profile to determine services to be provided;
        curr_state ← ls₁;
        provide services required in state curr_state;
        curr_state ← transition_to_next_state(curr_state,
                                   service_state_info);
        if (curr_state ∈ Pivot[])
            perform service state handoff; // Control back to
                                           // domain F
        endif
    end if;
    provide services required in state curr_state;
    curr_state ← transition_to_next_state(curr_state,
                                   service_state_info);
    if (curr_state ∈ Pivot[]) || curr_state == ls_n )
            perform service state handoff; // Control back
                                           // to domain F
    endif
```

Figure 5.3   The CMM/SS algorithm for the *L* domain.

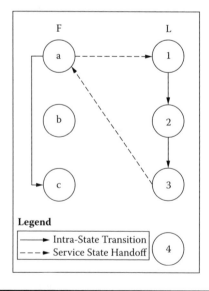

Figure 5.4   State transitions and service-state handoffs.

Because the services reside in L, they are executed in that domain. Their execution leads to state transitions local to L until a pivot state is reached where the service-state handoff will occur; then control will be passed back to F in the same state p from which the transition occurred (Figure 5.4). Along with passing the control, L also imparts enough information to F, allowing it to transition to a certain state $q \in F[s'] (p! = q$, and q may not be the next adjacent state after p). The choice of state q depends on the service execution logic. For instance, if the service logic in L decides to terminate the call, L will affect the appropriate state transition in F. Thus, both F and L maintain global state s^G of the call, which is updated after every service-state handoff. This synchronization is important because the call is actually serviced by two different signaling protocols. The global state s^G reflects the shared and authoritative state of the call.

Note that in Figure 5.4, the state labeled "b" in F appears not to be reachable. This is not an error, but in this particular example signifies that the service logic in L caused a state transition to occur to a nonadjacent state labeled "c" in F. Likewise, the state labeled "4" is not transitioned to in L; again, that is not an error, but instead signifies that the service completed in state 3, where a service-state handoff occurred.

5.3.4 CMM/SS: Issues

The core issue to consider in CMM/SS is this: What are the limitations of the mapping and service-state handoffs? Clearly, a mapping provides a way to abstract the transitions between L and F, and as such, there will be limitations. It is possible to encounter mutual state machines that cannot be integrated (essentially, where $\emptyset(x) = 0$, from Equation 5.4) or ones where the resultant mapped state machine is as complicated as the cross product of the individual state machines. The call models we analyzed in the telephony domain did not exhibit such behavior, and the mappings we accomplished through CMM/SS were not a complex cross product of the individual state machines.

To discuss the limitations, a mapping specified in Equation 5.5 is *complete* if there is not any semantic loss in service execution as a result of applying the CMM/SS technique. Likewise, a mapping is considered *partially complete* if there is a minimal semantic loss introduced as the result of applying the technique. It should be noted that unless two call models are exactly alike, all mappings will be partially complete. This is attributed to the fact that the design of two randomly picked call models is rarely alike in every respect; thus, mapping one to another necessarily introduces some semantic loss. This semantic loss is an outcome of applying information from protocol elements of F into those of L. However, so long as the mapping does not distort how the call model behaves in

L if the mapping was not applied in the first place, partially complete mappings are indeed the only logical outcome of the CMM/SS technique.

An example illustrates this: In the Originating Basic Call State Model (O_BCSM) of PSTN/IN, PIC 5 is used to select an outgoing circuit toward the destination endpoint (see Section 3.1.7). However, when an Internet telephony endpoint presents a SIP request to PSTN/IN for routing, selecting an outgoing circuit is of no help because the routing of the SIP request is performed on the Internet, not the PSTN. Instead, Internet-specific routing techniques need to be used to send the signaling message forward. This example captures our notion of partially complete mappings with minimal semantic loss: the end result is the same — i.e., a processing entity generates enough information to forward the signaling message; however, the means to get to the recipient address of the signaling message differ. Our goal is to enable a wide range of services while keeping the semantic loss to the bare minimum. If this can be done successfully, then the technique has proved its usefulness.

5.4 Implementing CMM/SS

An embryonic form of CMM/SS was first attempted in accessing the IN services from H.323 endpoints [CHI00]; we subsequently groomed the technique and formalized it through its application to accessing services from SIP endpoints as well [GUR02, GUR03d]. The results of that effort are discussed in this section.

Recall from Chapter 2 that services in the PSTN are provided by the IN. A PSTN switch, while processing a phone call, may temporarily suspend processing and consult the Service Control Point (SCP) on how to handle a particular call. The SCP executes the service and passes the results back to the switch in the form of instructions on the continuation of call processing. Thus, there is a close coupling between a switch and the SCP (or any other entity in the core providing value-added services) and all but the most basic of services are provided by the entities in the core of the network. In the Internet, by contrast, endpoints are themselves capable of executing complex services without the need for a centralized service platform. Thus, a preliminary step for implementing CMM/SS is to reconcile these two divergent views because the request for service will be placed by a SIP endpoint (and not a telephone endpoint connected to a traditional switch), while the service itself will be executed by traditional telephony equipment in the network core.

From the vantage point of the IN elements like the SCP, the fact that the request originated from a SIP entity versus a call processing function on a traditional switch is immaterial (assuming, of course, that the SCP does have access to the Internet, which all of them do because an SCP is nothing but a Sun Microsystems computer tuned to the telephony domain). It is also important that the SIP entity be able to provide features normally provided by the traditional switch, including interfacing with the IN network to access services. It should also maintain call state and trigger queries to the IN-based services, just as traditional switches do. Clearly, doing this in a SIP endpoint itself is not feasible because every SIP endpoint in existence would have to be upgraded to understand the IN call model and its interactions with the PSTN/IN for service signaling. Instead, a SIP intermediary, such as a proxy server or a Back-to-Back UA (B2BUA), may act as the functional equivalent of a traditional switch while processing a call from a SIP endpoint that requires access to the IN services. Generally speaking, proxy servers can be used for the IN services that occur during a call setup and teardown. For the IN services requiring specialized media handling (such as dual tone multifrequency (DTMF) detection) or specialized call control (such as placing parties on hold), B2BUAs will be required.

The most expeditious manner for providing existing IN services in IP is to use the deployed IN infrastructure as much as possible. The logical point in SIP to tap into for accessing existing IN services is an intermediary located physically closest to the SIP endpoint issuing the request (for originating IN services) or terminating the request (for terminating IN services). However, SIP intermediaries do not run an IN call model; to access the IN services transparently, the trick is to overlay the state machine of the SIP entity with an IN layer such that call acceptance and routing is performed by the native SIP state machine and services are accessed through the IN layer using an IN call model. Such an IN-enabled SIP intermediary, operating in synchrony with the events occurring at the SIP transaction level and interacting with the IN elements, is depicted in Figure 5.5.

The SIP intermediary of Figure 5.5, which we will refer to as a CMM/SS entity in the rest of the chapter, accepts a session setup request and processes it initially using the normal SIP state machines. However, at certain pivot states, a service-state handoff occurs to the IN layer, which performs further processing by interfacing with the PSTN/IN layer. The list of pivot states for SIP and its mapping into PSTN/IN Q.1204 BCSM will be detailed in Section 5.4.4.

5.4.1 CMM/SS Considerations

When interworking between Internet telephony and PSTN/IN networks, the main issue is to translate between the states produced by the Internet telephony signaling and those used in traditional IN environments. Such a translation entails attention to the considerations listed below.

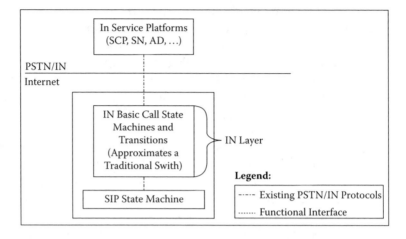

Figure 5.5   A CMM/SS entity.

5.4.1.1   The Concept of a Call State Model in SIP

The concept of a call state is porous in SIP; SIP is a transaction-stateful protocol. The IN services occur within the context of a call, i.e., during call setup, teardown, or in the middle of a call. SIP entities such as proxies, where some of these services may be realized, typically run in a transaction-stateful (or stateless) mode. In such a mode, a SIP proxy that handled the initial INVITE is not guaranteed to receive a subsequent request, such as a BYE. Fortunately, SIP has primitives to force proxies to run in a call-stateful mode, namely, the Record-Route header. This header forces the user agent client (UAC) and user agent server (UAS) to create a route set, which consists of all intervening proxies through which subsequent requests must traverse. Thus, SIP proxies must run in a call-stateful mode to provide the IN services on behalf of the user agents.

A B2BUA is another SIP element where the IN services can be realized. Because a B2BUA is a true SIP UA, it maintains the complete call state and is thus capable of providing the IN services as a first-class citizen of the signaling ecosystem.

Natively, SIP maintains a transaction state, in lieu of an overall call state. The SIP specification contains detailed state models for an INVITE transaction and a non-INVITE transaction from the viewpoint of a UAS and a UAC (Figures 5 to 8 in [ROS02]). However, it does not contain an aggregate figure for an overall call model from a call initiation to termination. Harvested from Figures 5 to 8 of [ROS02], we present an aggregate SIP call model in Figure 5.6.

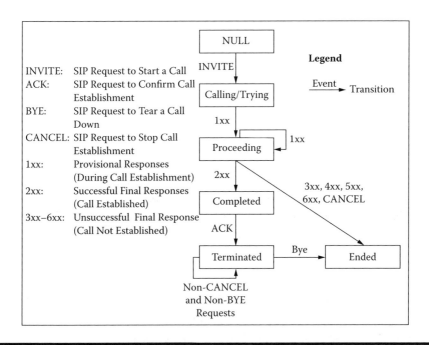

Figure 5.6 An aggregate SIP state machine.

Compared to the Q.1204 IN BCSM, the SIP state machine of Figure 5.6 is extremely simple. Unlike its Q.1204 counterpart, it does not have two explicit halves. Instead, the same state machine represents both halves of the call, or in SIP parlance, the UAC or UAS dictates the context under which the state machine is operating; the number of states remain the same.

The SIP state machine depicted above contains six states and eight transitions. The "Terminated" state is entered through an ACK for a 2xx-class response, because, in this case, the ACK is considered a separate transaction in SIP. For other responses, the ACK is part of the INVITE transaction, and we have chosen not to explicitly model it with a state. In Figure 5.6, we have chosen to stay true to the names of the states in Figures 5 to 8 of [ROS02]; thus, it seems somewhat incongruous to transition from "Terminated" to "Ended," but we feel that resemblance of state names between Figure 5.6 and Figures 5 to 8 of [ROS02] would aid the reader in relating the individual SIP state machines of [ROS02] to the aggregated one presented above.

We will use the aggregate SIP state machine of Figure 5.6 when we demonstrate the mapping between a SIP state machine and PSTN/IN Q.1204 BCSM in later sections.

5.4.1.2 Relationship between an SCP and a CMM/SS Entity

In the architecture model we propose, each CMM/SS entity is preconfigured to communicate with one logical SCP server, using whatever communication mechanism is appropriate. Different SIP servers (e.g., those in different administrative domains) may communicate with different SCP servers, so that there is no single SCP server responsible for all SIP servers.

As Figure 5.5 depicts, the IN portion of the CMM/SS entity will communicate with the SCP. This interface between the IN call-handling layer and the SCP is an implementation decision and, indeed, can be any one of the following depending on the interfaces supported by the SCP: Intelligent Network Application Part (INAP) (or Transaction Capabilities Application Part (TCAP)) over IP, INAP (or TCAP) over SIGTRAN, or INAP (or TCAP) over Signaling System 7 (SS7).

5.4.1.3 Support of Announcements and Mid-Call Signaling

Services in the IN such as credit card calling typically play announcements and collect digits from the caller before a call is set up. Playing announcements and collecting digits require the manipulation of media streams. In SIP, proxies do not have access to the media data path. Thus, such services should be executed in a B2BUA.

Although the SIP specification [ROS02] allows for endpoints to be put on hold during a call, or a change of media streams to take place, it does not have any primitives to transport other mid-call control information. This may include transporting DTMF digits, for example. Extensions to SIP, such as the INFO method [DON00] or the SIP event notification extension [ROA02], can be considered for services requiring mid-call signaling. Alternatively, DTMF can be transported in Real-Time Transport Protocol (RTP) itself [SCH03].

5.4.2 CMM/SS Architectural Model

Figure 5.7 depicts an architectural model for the IN service control based on our approach. On both the originating and terminating sides, a CMM/SS entity is assumed to be present (it could be a proxy or a B2BUA). In the figure, we implicitly assume that one of the two endpoints involved in a session is on the PSTN, but this need not be the case. We have done so to provide a context for understanding the workings of CMM/SS. CMM/SS does, however, require that at least one endpoint be on the Internet because the call request will originate (or terminate) on that endpoint. If

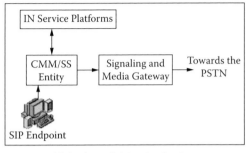

(a) Applying Originating Services

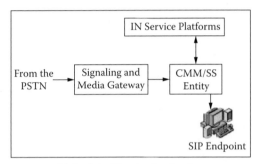

(b) Applying Terminating Services

Figure 5.7   Applying IN services to SIP endpoints.

both endpoints reside on the Internet, then the PSTN is used simply to access the service (which resides in the PSTN domain), not to route the call request or provide media capabilities.

Figure 5.7a shows originating-side services being applied to the SIP endpoint. When the CMM/SS entity receives the call from an endpoint in its domain, it performs a service-state handoff to the IN layer for subsequent processing and awaits further instructions. The IN layer applies services appropriate to the originating half of the Q.1204 BCSM, or O_BCSM. Conversely, Figure 5.7b demonstrates a CMM/SS entity receiving requests from the PSTN and applying services appropriate to the terminating half of the Q.1204 BCSM, or T_BCSM, to the SIP endpoint.

5.4.3   Realizing CMM/SS in Software

We have authored two pieces of software to demonstrate CMM/SS; both pieces taken together essentially represent the CMM/SS entity depicted in Figure 5.5.

The first component we authored was a PSTN/IN call model written in C++. This call model reproduces the state machines for the originating and terminating halves of the Q.1204 BCSM, including all legal transitions between them. This software acted as an IN layer between a SIP proxy and the PSTN service platform (see Figure 5.5). The next piece of software we authored was an RFC3261-compliant SIP proxy server that implemented the state machines of Figures 5 to 8 of [ROS02], as well as the aggregated state machine depicted in Figure 5.6. The SIP proxy server contained hooks into the IN layer, which was mapped to the SIP state machine (the mapping itself will be discussed in Section 5.4.4). When the SIP proxy server received a request from the network, it initialized the O_BCSM object in the IN layer, which would, in turn, interface with the PSTN service to inform the proxy on the treatment to be applied to a session setup request. Likewise, when the SIP proxy was ready to send the session setup request downstream (i.e., toward a UAS), it would initialize the T_BCSM object in the IN layer and apply terminating services to the session setup request. In this manner, IN services were provided to Internet endpoints in a transparent fashion; the endpoints were not cognizant of the fact that the PSTN was providing them critical services.

The CMM/SS entity maintained a global state, s^G, in the form of the data structure presented in Figure 5.8. This data structure was initialized by the CMM/SS entity and passed to the IN layer. The IN layer then took over and, depending on the current call state, provided appropriate services at each PIC. Control of the session logic toggled between the proxy and the IN layer, each applying appropriate processing to it. The proxy was ultimately responsible for delivering (and routing) the call, while the IN layer was responsible for providing services.

```
typedef struct call_info{
    char CRV[CRV_SZ]; // Transaction ID
    char CdPN[NUM_SZ]; // Called Party Number
    char CgPN[NUM_SZ]; // Calling Party Number
    int  Current_State; // For IN Call Model
    int  Suggested_Next_State; // For IN Call
                               // Model
    unsigned long DP_OBCSM; // DPs for the O_BCSM
    unsigned long DP_TBCSM; // DPs for T_BCSM
    long start_time; // for billing
    long stop_time; // for billing
};
```

Figure 5.8   Shared-state data structure.

5.4.4 Applying the Mapping

To apply the mapping between the SIP state machine and Q.1204 BCSM, we followed the CMM/SS technique and algorithms listed in Section 5.3. The SIP state machine corresponds to the F domain and the Q.1204 BCSM corresponds to the L domain. The states of $F[s]$ need to be mapped into L such that we satisfy Equation 5.4. Because F and L contain a different number of states — the Q.1204 PSTN/IN call model consists of 19 states and 35 transitions (11 states and 21 transitions in the originating BCSM, and 8 states and 14 transitions in the terminating one), and the SIP state machine of Figure 5.6 contains 6 states and 8 transitions — there will not be a one-to-one mapping between states.

We now present the mapping from SIP to O_BCSM and T_BCSM, respectively. In the mapping below, our reference to a particular SIP state is in relation to the states listed in Figure 5.6.

5.4.4.1 Mapping SIP to O_BCSM

To map the SIP state machine to O_BCSM, we followed the CMM/SS technique of aligning the two call models on the two yellow bulbs: Calling/Trying for SIP, and O_NULL for O_BCSM. We then established pivot states. For SIP, the set of pivot states consists of

Pivot = {Calling/Trying, Proceeding, Terminated, Ended}

For O_BCSM, the set of pivot states consists of

Pivot = {O_NULL, O_Exception, Call_Sent, O_Active, O_Disconnect}

The 11 PICs of O_BCSM come into play when a call request (SIP INVITE message) arrives from an upstream SIP client to an originating CMM/SS entity running the IN call model. This entity will create an O_BCSM object and initialize it in the O_NULL PIC. The next seven IN PICs — O_NULL, AUTH_ORIG_ATT, COLLECT_INFO, ANALYZE_INFO, SELECT_ROUTE, AUTH_CALL_SETUP, and CALL_SENT — can all be mapped to the SIP Calling/Trying state.

Figure 5.9 provides a visual mapping from the SIP state machine to the originating half of the IN call model. Note that the service-state handoffs occur at appropriate times, resulting in control of the session setup shuttling between the SIP machine and the IN O_BCSM call model. The SIP Calling/Trying state has enough functionality to absorb the seven PICs as described below:

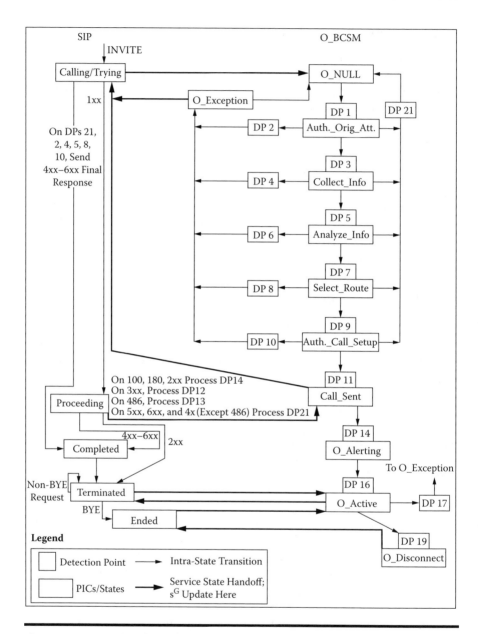

Figure 5.9   Mapping from SIP to O_BCSM.

O_NULL: This PIC is basically a fall-through state to the next PIC, AUTHORIZE_ORIGINATION_ATTEMPT.

AUTHORIZE_ORIGINATION_ATTEMPT: In this PIC, the IN layer has detected that someone wishes to make a call. Under some circumstances (e.g., the user is not allowed to make calls during certain hours), such a call cannot be placed. SIP has the ability to authorize the calling party using a set of policy directives configured by the SIP administrator. If the called party is authorized to place the call, the IN layer is instructed to enter the next PIC, COLLECT_INFO, through DP 3 (Origination_Attempt_Authorized). If for some reason the call cannot be authorized, DP 2 (Origination_Denied) is processed and control transfers to the SIP state machine. The SIP state machine must format and send a non-2xx final response (possibly 403) to the UAC.

COLLECT_INFO: This PIC is responsible for collecting a dial string from the calling party and verifying the format of the string. If overlap dialing is used, this PIC can invoke DP 4 (Collect_Timeout) and transfer control to the SIP state machine, which will format and send a non-2xx final response (possibly a 484). If the dial string is valid, DP 5 (Collected_Info) is processed and the IN layer is instructed to enter the next PIC, ANALYZE_INFO.

ANALYZE_INFO: This PIC is responsible for translating the dial string to a routing number. Many IN services such as freephone (800 number), Local Number Portability (LNP), Originating Call Screening (OCS), etc., occur during this PIC. The IN layer can use the Request-Uniform Resource Identifier (R-URI) of the SIP INVITE request for analysis. If the analysis succeeds, the IN layer is instructed to enter the next PIC, SELECT_ROUTE. If the analysis fails, DP 6 (Invalid_Info) is processed and control transfers to the SIP state machine, which will generate a non-2xx final response (possibly one of 400, 401, 403, 404, 405, 406, 410, 414, 415, 416, 485, or 488) and send it to the upstream entity.

SELECT_ROUTE: In the circuit-switched network, the actual physical route has to be selected at this point. The SIP analog of this would be to determine the next-hop SIP server. The next-hop SIP server could be chosen by a variety of means. For instance, if the R-URI in the incoming INVITE request is an E.164 number, the SIP entity can use a protocol like Telephony Routing over Internet Protocol (TRIP) [ROS02a] to find the best gateway to egress the request onto the PSTN. If a successful route is selected, the IN call model moves to PIC AUTH_CALL_SETUP via DP 9 (Route_Selected). Otherwise, the control transfers to the SIP state machine via DP 8 (Route_Select_Failure), which will generate a non-2xx final response (possibly 488) and send it to the UAC.

AUTH_CALL_SETUP: Certain service features restrict the type of call that may originate on a given line or trunk. This PIC is the point at which relevant restrictions are examined. If no such restrictions are encountered, the IN call model moves to PIC CALL_SENT via DP 11

(Origination_Authorized). If a restriction is encountered that prohibits further processing of the call, DP 10 (Authorization_Failure) is processed and control is transferred to the SIP state machine, which will generate a non-2xx final response (possibly 404, 488, or 502). Otherwise, DP 11 (Origination_Authorized) is processed and the IN layer is instructed to enter the next PIC, CALL_SENT.

CALL_SENT: At this point, the request needs to be sent to the downstream entity, and the IN layer waits for a signal confirming either that the call has been presented to the called party or that a called party cannot be reached for a particular reason. The control is now transferred to the SIP state machine. The SIP state machine should now send the call to the next downstream server determined in PIC SELECT_ROUTE.

If the above seven PICs have been successfully negotiated, the CMM/SS entity now sends the SIP INVITE message to the next-hop server. Further processing now depends on the provisional responses (if any) and the final response received by the SIP state machine. The core SIP specification does not guarantee the delivery of 1xx responses; thus, special processing is needed at the IN layer to transition to the next PIC (O_ALERTING) from the CALL_SENT PIC. The special processing needed for responses while the SIP state machine is in the "Proceeding" state and the IN layer is in the CALL_SENT state is described next:

A 100 response received at the SIP state machine elicits no special behavior in the IN layer.

A 180 response received at the SIP entity enables the processing of DP 14 (O_Term_Seized); however, a state transition to O_ALERTING is not undertaken yet. Instead, the IN layer is instructed to remain in the CALL_SENT PIC until a final response is received.

A 2xx response received at the SIP entity enables the processing of DP 14 (O_Term_Seized) and the immediate transition to the next state, O_ALERTING (processing in O_ALERTING is described later).

A 3xx response received at the CMM/SS entity enables the processing of DP 12 (Route_Failure). The IN call model from this point goes back to the SELECT_ROUTE PIC to select a new route for the contacts in the 3xx final response (not shown in Figure 5.9 for brevity).

A 486 (Busy Here) response received at the CMM/SS entity enables the processing of DP 13 (O_Called_Party_Busy), and resources for the call are released at the IN call model.

If the CMM/SS entity gets a 4xx (except 486), 5xx, or 6xx final response, DP 21 (O_Calling_Party_Disconnect_&_O_Abandon) is processed and control passes to the SIP state machine. Because a call was not successfully established, both the IN layer and the SIP state machine can release resources for the call.

O_ALERTING: This PIC will be entered as a result of receiving a 200-class response. Because a 200-class response to an INVITE indicates acceptance, this PIC is mostly a fall through to the next PIC, O_ACTIVE via DP 16 (O_Answer).

O_ACTIVE: At this point, the call is active. Once in this state, the call may get disconnected only when one of the following three events occurs: (1) the network connection fails, (2) the called party disconnects the call, or (3) the calling party disconnects the call. If event 1 occurs, DP 17 (O_Connection_Failure) is processed and call control is transferred to the SIP state machine. Because the network failed, there is not much sense in attempting to send a BYE request; thus, both the SIP state machine and the IN call layer should release all resources associated with the call and initialize themselves to the null state. The occurrence of event 2 results in the processing of DP 19 (O_DISCONNECT) and a move to the last PIC, O_DISCONNECT. Event 3 would be caused by the calling party proactively terminating the call. In this case, DP 21 (O_Abandon_&_O_Calling_Party_Disconnect) will be processed and control passed to the SIP state machine. The SIP state machine must send a BYE request and wait for a final response. The IN layer releases all its resources and initializes itself to the null state.

A salient point about PIC O_ACTIVE is that all mid-call SIP-related signaling arriving at the CMM/SS entity forces a service-state handoff to this IN state. The IN BCSM can apply the appropriate mid-call service treatment to the session and execute a service-state handoff back to the IN layer.

O_DISCONNECT: When the SIP entity gets a BYE request, the IN layer is instructed to move to the last PIC, O_DISCONNECT, via DP 19. A final response for the BYE is generated and transmitted by the CMM/SS entity, and the call resources are de-allocated by the SIP state machine as well as the IN layer.

5.4.4.2   Mapping SIP to T_BCSM

To map the SIP state machine to T_BCSM, we followed the CMM/SS technique of aligning the call models on the two yellow bulbs: "Proceeding" for SIP, and T_NULL for T_BCSM. As before, we then established pivot states. For SIP, the set of pivot states consists of

$$\text{Pivot} = \{\text{Proceeding, Terminated, Ended}\}$$

For O_BCSM, the set of pivot states consists of

Pivot = {T_NULL, T_Exception, T_Active, T_Disconnect}

The T_BCSM object is created when a SIP INVITE message makes its way to the terminating CMM/SS entity, which creates the T_BCSM object and initializes it to the T_NULL PIC. The mapping of 8 states and 14 transitions of the terminating half of the Q.1204 BCSM into an equivalent SIP state machine is reproduced in Figure 5.10.

The SIP "Proceeding" state has enough functionality to absorb the first five PICS — T_Null, Authorize_Termination_Attempt, Select_Facility, Present_Call, T_Alerting — as described below:

> T_NULL: At this PIC, the terminating end creates the call at the IN layer. The incoming call results in the processing of DP 22, Termination_Attempt, and a transition to the next PIC, AUTHORIZE_TERMINATION_ATTEMPT, takes place.
>
> AUTHORIZE_TERMINATION_ATTEMPT: In this PIC, the fact that the called party wishes to receive the call is ascertained and that the facilities of the called party are compatible with those of the calling party. If any of these conditions is not met, DP 23 (Termination_Denied) is invoked and the call control is transferred to the SIP state machine. The SIP state machine can format and send a non-2xx final response (possibly 403, 405, 415, or 480). If the conditions of the PIC are met, processing of DP 24 (Termination_Authorized) is invoked and a transition to the next PIC, SELECT_FACILITY, takes place.
>
> SELECT_FACILITY: The intent of this PIC in circuit-switched networks is to select a line or trunk to reach the called party. Because lines or trunks are not applicable in an IP network, a CMM/SS entity can use this PIC to interface with a PSTN gateway and select a line or trunk to route the call. If the called party is busy, or a line or trunk cannot be seized, the processing of DP 25 (T_Called_Party_Busy) is invoked, followed by a transition of the call to the SIP state machine. The SIP state machine must format and send a non-2xx final response (possibly 486 or 600). If a line or trunk was successfully seized, the processing of DP 26 (Terminating_Resource_Available) is invoked and a transition to the next PIC, PRESENT_CALL, takes place.
>
> PRESENT_CALL: At this point, the call is presented (via an appropriate PSTN signaling protocol such as the ISDN User Part (ISUP) Association for Computing Machinery (ACM) message or Q.931 alerting message, or simply by ringing a PSTN phone). If there was an error presenting

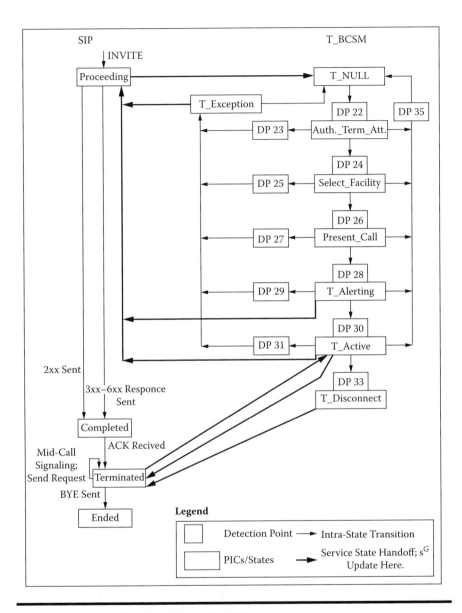

Figure 5.10   Mapping from SIP to T_BCSM.

the call, the processing of DP 27 (Presentation_Failure) is invoked and the call control is transferred to the SIP state machine. The SIP state machine must format and send a non-2xx final response (possibly 480). If the call was successfully presented, the processing of DP 28 (T_Term_Seized) is invoked and a transition to the next PIC, T_ALERTING, takes place.

T_ALERTING: At this point, the called party is alerted. Control is now passed momentarily to the SIP state machine, so it can generate and send a "180 Ringing" response to its peer. Furthermore, because network resources have been allocated for the call, timers are set to prevent indefinite holding of such resources. The expiration of the relevant timers results in the processing of DP 29 (T_No_Answer) and the call control is transferred to the SIP state machine. The SIP state machine must format and send a non-2xx final response (possibly 408). If the called party answers, then DP 30 (T_Answer) is processed, followed by a transition to the next PIC, T_ACTIVE.

The rest of the PICs after the above five have been negotiated are mapped as follows:

T_ACTIVE: The call is now active. Once this state is reached, the call may become inactive only under one of the following three conditions: (1) the network fails the connection, (2) the called party disconnects the call, or (3) the calling party disconnects the call. Event 1 results in the processing of DP 31 (T_Connection_Failure) and call control is transferred to the SIP state machine. Because the network failed, there is not much sense in attempting to send a BYE request; thus, both the SIP state machine and the IN call layer should release all resources associated with the call and initialize themselves to the null state. Event 2 results in the processing of DP 33 (T_Disconnect) and a transition to the next PIC, T_DISCONNECT. Event 3 would be caused by the receipt of a BYE request at the SIP state machine. Resources for the call should be de-allocated and the SIP state machine must send a 200 OK for the BYE request (not shown in Figure 5.10).

A salient point about T_ACTIVE PIC is the treatment of mid-call signaling. Once the session has been established, an IN service may perform mid-call signaling. If this happens, a service-state transfer occurs to the SIP "Terminated" state and a SIP method appropriate to the mid-call signaling is sent out. Upon receipt of a response, another service-state transfer will occur, putting the control back in the IN layer.

T_DISCONNECT: In this PIC, the disconnect treatment associated with the called party's having disconnected the call is performed at the IN layer. A service-state transfer occurs to the SIP "Terminated" state with enough information passed in s^G to aid the SIP state machine in sending a BYE request out.

As part of the mapping of the SIP state machine to the two halves of Q.1204 BCSM, Figure 5.9 and Figure 5.10 indicate a relation between the

Table 5.1   Correlating SIP Response Codes with DPs

SIP Response Code	IN DP
200 OK	DP 14
3xx Redirection	DP 12
403 Forbidden	DP 2, DP 21, DP 23
484 Address Incomplete	DP 4, DP 21
400 Bad Request	DP 6, DP 21
401 Unauthorized	DP 6, DP 21
404 Not Found	DP 6, DP 10, DP 21
405 Method Not Allowed	DP 6, DP 21, DP 23
406 Not Acceptable	DP 6, DP 21
408 Request Timeout	DP 29
410 Gone	DP 6, DP 21
414 Request-URI Too Long	DP 6, DP 21
415 Unsupported Media Types	DP 6, DP 21, DP 23
416 Unsupported URI Scheme	DP 6, DP 21
480 Temporarily Unavailable	DP 23, DP 27
485 Ambiguous	DP 6, DP 21
486 Busy Here	DP 13, DP 21, DP 25
488 Not Acceptable Here	DP 6, DP 8, DP 10, DP 21
502 Bad Gateway	DP 10, DP 21
600 Busy Everywhere	DP 21, DP 25

DPs and SIP response codes. The processing of a certain DP may result in the SIP state machine sending out an appropriate SIP response. Table 5.1 contains a mapping of SIP responses (2xx to 6xx) to their appropriate DPs.

Our work in call model mapping between SIP and the two halves of Q.1204 BCSM outlined in Figure 5.9 and Figure 5.10 has been a subject of an Internet Engineering Task Force (IETF) Informational Request for Comment (RFC) document [GUR02]. The RFC series is the official publication channel for Internet standards documents and other publications of the Internet community [BRA96]. As of this writing, [GUR02] has been peer-reviewed by the SIP-related working groups in the IETF and is in the RFC editor's queue, waiting on an assignment of an RFC number for final publication.

5.5   Results from CMM/SS

To experimentally prove the feasibility of CMM/SS, we attempted four benchmark services. These services were chosen as a mix of origination and termination BCSM services. Two services were drawn from the O_BCSM

Table 5.2   Benchmark Services Accomplished in CMM/SS

Service Name	BCSM Half	DP Involved
Originating Call Screening (OCS)	O_BCSM	DP 5
Abbreviated dialing (AD)	O_BCSM	DP 7
Call forwarding (CF)	T_BCSM	DP 22
Calling Name Delivery (CNAM)	T_BCSM	DP 22

half of the IN call model, and two were drawn from the T_BCSM half. The services depend entirely on signaling for their execution, i.e., they do not involve any media components (tone detection, for example). Table 5.2 contains the benchmark services realized through CMM/SS, in which half of the BCSM they occur, and which DPs are involved in the service.

5.5.1 Network Topology

Our laboratory setup consisted of several SIP endpoints (each running a SIP user agent client and a SIP user agent server), a SIP proxy server fortified with the PSTN/IN call layer (the CMM/SS entity), and an SCP execution environment that serviced requests. The SCP execution environment hosted the PSTN service and executed it in response to requests arriving at it from a telephony switch. In our case, the CMM/SS entity acted as a telephony switch by sending it PSTN service requests. Figure 5.11 depicts this setup.

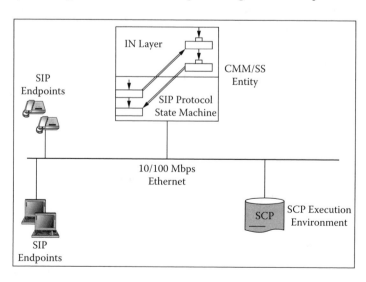

Figure 5.11   Network topology.

Each SIP UA was configured, upon boot-up, to register with the CMM/SS entity using a telephone number. This is important — the intent is to mimic PSTN/IN services offered on the PSTN; hence, endpoints are identified using telephone numbers and not the more powerful and generic e-mail-like SIP URI. The SCP execution environment was configured with the data and service logic pertaining to the four benchmark IN services. Communications between CMM/SS and the SCP utilized the Ethernet network. Whenever the CMM/SS wanted to execute a service in the SCP execution environment, it would format a TCAP message and transmit it over the IP network. The SCP would execute the service and the response would arrive back to the CMM/SS over the IP network.

5.5.2 Results

The results obtained from the implementation validate the CMM/SS technique. We summarize them in Table 5.3. The second column contains provisioning information required for the service to operate (the data). The third column contains the behavior of the service in the PSTN, and the fourth column details the behavior of the service with the application of CMM/SS. The behavior in the fourth column is almost identical to that of the third column. This provides an empirical proof of validation of the CMM/SS technique.

Table 5.3 CMM/SS Results

Service	Configuration Data	Behavior with Native PSTN	Behavior with CMM/SS
OCS	A blocked number list	Caller hears fast busy tone	Call request rejected with 403 Forbidden; some SIP user agents played a fast busy signal on receipt of a 403
AD	Telephone numbers	The abbreviated number is expanded and routed to its destination	The PSTN/IN call layer returns a translated URI to which the SIP proxy routes the call
CNAM	PSTN name database	Callee's name is displayed in a caller ID device	Callee's name is displayed in the SIP UA graphical user interface (GUI)
CF	Telephone numbers	Incoming call is forwarded to a new destination	The PSTN/IN call layer returns a translated URI to which the SIP proxy routes the call

As can be observed from Table 5.3, the behavior of the service with CMM/SS is similar to the behavior of the service when executed natively in the PSTN. It is extremely important to state that the service logic running on the SCP execution environment and the protocol required to access the SCP were assumed immutable; neither the logic nor the protocol was modified to account for the fact that the service request was now being sent by a SIP entity, and not a PSTN switch. In all cases, the service executed flawlessly on the SCP execution environment, impervious to the fact that a vastly different protocol was used by the endpoint involved in the session setup or teardown.

5.5.3 Service Description and Call Flows

For each service we realized through CMM/SS, we now present a detailed description and a relevant call flow. Note that in the call flow, SIP messages are reproduced in an abbreviated form for brevity, i.e., not all headers and bodies are shown.

5.5.3.1 Originating Call Screening (OCS)

OCS is a service whereby the O_BCSM ensures that the caller is authorized to initiate a call to the dialed number (or the callee). The OCS service is accessed by arming the Collect_Info trigger (DP 5) of the O_BCSM of the IN call model. When the CMM/SS entity receives an INVITE request, it extracts the Request-URI (an E.164 number) of the party invited and sends it, along with other information, to the portable IN call layer. The IN call layer proceeds through its PICs and, on reaching an armed DP 5, triggers a TCAP request to the SCP. The SCP analyzes this request and instructs the CMM/SS entity on what to do next with the call. The SCP has access to a user profile database, which contains, among other fields, a column that restricts the caller from making certain calls (for example, 900-number calls, which in the United States are billed at higher than normal rates — hence the need to restrict such calls).

In the call flow examples below, the letters C, S, and N are used to identify the SIP user agent server (caller), the CMM/SS entity, and the next-hop SIP server (UAS, or another proxy, or a gateway), respectively.

Example: C Wants to Initiate a Call to a 900 Number

```
C->S: INVITE sip:9005551111@service-provider.com;user=phone
      From: "Vijay K. Gurbani" <sip:vkg@iit.edu>;tag=as-909dd-fe
      To: sip:9005551111@ service-provider.com
```

```
Via: SIP/2.0/UDP temphost1.iit.edu;branch=z9hG4bK91
Call-ID: CC6677901AF@temphost1.iit.edu
CSeq: 1 INVITE
...
```

S extracts the SIP Request-URI (sip:9005551111@service-provider.com) and the values of the To: and From: fields and sends them to the IN call layer. The IN call layer formats a TCAP request, sends it to the SCP, and is told that the caller does not have sufficient privileges to continue with the call. S sends the following SIP (final) response to C:

```
S->C: SIP/2.0 403 Forbidden
      From: "Vijay K. Gurbani" <sip:vkg@iit.edu>;tag=as-909dd-fe
      To: sip:9005551111@ service-provider.com;tag=0233-112322a66
      Via: SIP/2.0/UDP temphost1.iit.edu;branch=z9hG4bK91
      Call-ID: CC6677901AF@temphost1.iit.edu
      CSeq: 1 INVITE
      Content-Length: 0
```

5.5.3.2   Abbreviated Dialing (AD)

AD is a service feature that permits the caller to dial fewer digits than are required under a national numbering plan to access the PSTN. It has also been used to implement network-based speed dialing, a feature sold by telephone service providers that allows users to dial a fewer number of digits to initiate a call. The network fills in the missing digits by expanding on a subset of digits presented to it. Service is accessed by arming the Analyse_Info trigger (DP 7) of the O_BCSM of the IN call model.

Example: C Calls a Number Using AD

```
C->S: INVITE sip:73000@iit.edu;user=phone
      From: "Vijay K. Gurbani" <sip:vkg@iit.edu>;tag=as-909dd-fe
      To: sip:9005551111@iit.edu
      Via: SIP/2.0/UDP temphost1.iit.edu;branch=z9hG4bK91
      Call-ID: CC6677901AF@temphost1.iit.edu
      CSeq: 1 INVITE
      ...
```

When S receives an INVITE request, it extracts the Request-URI (an E.164 number) of the party invited and sends it, along with other information, to the IN call layer. The IN call layer proceeds through its PICs and, on reaching an armed DP 7, triggers a TCAP request to the SCP execution environment. The SCP analyzes this request and, after consulting an AD database, returns the new routing number to S (1-312-567-3000). S then forwards the request to the next-hop SIP server, after modifying the Request-URI to include the new routing number:

```
S->N: INVITE sip:13215673000@iit.edu;user=phone
      From: "Vijay K. Gurbani" <sip:vkg@iit.edu>;tag=as-909dd-fe
      To: sip:9005551111@iit.edu
      Via: SIP/2.0/SCTP border-host.iit.edu;branch=z9hG4bKkjshdyff
      Via: SIP/2.0/UDP temphost1.iit.edu;branch=z9hG4bK91
      Call-ID: CC6677901AF@temphost1.iit.edu
      CSeq: 1 INVITE
      ...
```

5.5.3.3   Call Forwarding (CF)

CF is another well-known service whereby the incoming phone call to the callee is forwarded to another number. Unlike the previous two services, this is a terminating-side service. This service is accessed by dynamically arming the Termination_Attempt trigger (DP 22) of the T_BCSM in the IN call model. The SIP call messages for this service are similar to those of AD, with the only difference being that the CMM/SS entity where this processing occurs is on the terminating side of the call.

5.5.3.4   Calling Name Delivery (CNAM)

CNAM is another well-known service sold commercially by the telephone service providers under the name caller ID. This service displays, for the called party, the name and number of the calling party. The service is a terminating-side service and is accessed by arming the Termination_Attempt trigger (DP 22) of the T_BCSM in the IN call model.

Interestingly enough, in SIP, this service could turn out to be simplistic if the "From" header of the SIP INVITE message contains the display name (a display name is an information element in SIP that lists the name of the person associated with a URI; for example, in the URI "Vijay K. Gurbani

<sip:vkg@iit.edu>," the display name consists of "Vijay K. Gurbani"). If the display name is absent, then the IN call layer uses the E.164 address to perform a TCAP query against the subscriber database to retrieve this information. Once a response is received, it is presented to the calling party using an appropriate display device.

5.6 Performance of CMM/SS

To characterize the performance of CMM/SS, we analyzed the behavior of a representative service logic running in the SCP execution environment. We chose the CNAM service, which is a terminating-side service and works by querying the IN databases for a display name, given a phone number. This sort of queryñresponse behavior is endemic to many IN services; thus, the CNAM service is a good representative of a class of IN services that perform similar functions.

Performance analysis was divided into two parts: First, we studied the behavior of the CNAM service operating under a traditional telephony environment. This included a CNAM service instance executing on the SCP execution environment and call requests originating from a SS7-based call simulator. All signaling was PSTN based, i.e., SIP endpoints were not involved at all. This analysis provided a baseline profile that would be used to compare the performance obtained through the CMM/SS technique.

Next, we studied the behavior of the service as it executes in the SCP execution environment, but the trigger for service execution occurred in a CMM/SS entity in lieu of an SS7-based call simulator. The SIP endpoint conversed with a CMM/SS entity, which, in turn, involved the service at the appropriate time. This analysis provided a target profile, which we compared to the baseline profile.

For the baseline profile, the performance tests were executed with the SCP execution environment running on a Sun Microsystems Netra 1400 with four processors and 4 Gbyte of memory. The SS7 signaling simulator was a commercially supplied product running on a separate host and communicating with the SCP execution environment over the local area network. For the target profile, the performance tests were executed with the UAC and UAS running on a similarly equipped Sun Microsystems Netra 1400. The CMM/SS and SCP execution environments were co-resident on another similarly equipped Sun Microsystems machine. Communications between the user agents and CMM/SS occurred over the local area network, and those between the CMM/SS and SCP execution environments transpired on the loopback interface.

The results of the measurements are presented in Table 5.4. The quantity of interest measured was the delay time it took the CNAM service

Table 5.4 Performance Results of CMM/SS

Delay (ms)	Baseline Profile	Target Profile	Difference
Mean (ms)	250.52	297.11	46.59 (18.6%)
Maximum (ms)	500.12	512.92	—
Minimum (ms)	102.28	97.22	—

to return a response, once a request was received by it. The sample size of the request transmitted was 25,000 requests over the course of the run. The delay time is characterized in the table for both the baseline and the target profile.

As can be observed, using CMM/SS introduces a delay of about 18.6 percent. This delay is attributed to two aspects: the introduction of CMM/SS technique and the SIP itself. Following the CMM/SS technique, an incoming request is actually treated by two call models: the SIP state machine and the IN T_BCSM. This straddled processing introduces some delay as service-state handoffs occur between the call models.

Additional delay is also introduced by the SIP itself. SIP is a textually oriented protocol, as such parsing and serialization of SIP takes more time when compared to binary epresentation protocols like SS7 [COR04]. For the CNAM service, the total delay under the CMM/SS model can be characterized by the following equation:

$$D = P_{uac} + \sum_{i=1}^{n} P_n + P_{cmm} + S_e + P_{uas} \qquad (5.6)$$

where total delay, D, is:

P_{uac} = SIP processing delay introduced by a UAC (constructing a SIP request, serializing it, and transmitting it).

P_n = SIP processing delay introduced by the nth intermediary. (A SIP proxy, for instance, will get the request, parse it, analyze it, and, subsequently, serialize it for sending it downstream.)

P_{cmm} = SIP processing delay introduced by the CMM/SS entity as the technique is applied to access the services.

S_e = Service execution time in the SCP execution environment.

P_{uas} = SIP processing delay introduced by a UAS (receiving the request and issuing a response).

In our benchmark, we set the summation of P_n to 0. This was because there were not any intermediaries between the UAC and UAS (aside from the CMM/SS entity); see the network configuration of Figure 5.11. Furthermore,

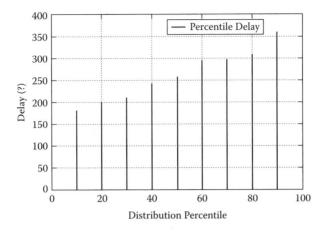

Figure 5.12   CMM/SS distribution percentiles.

because the service execution time, S_e, is the same between the baseline profile and the target profile, we can safely set it to a constant.

Thus, the components contributing to the 18.6 percent difference in the delay between the baseline profile and the target profile are P_{uac}, P_{cmm}, and P_{uas}. Of the 25,000 runs, each run was quantified by a delay that was measured as follows: we measured the time a request was sent from the UAC and the time it arrived at the UAS. In between, it traversed the CMM/SS, which applied the service treatment to the request by accessing the CNAM service in the PSTN. The difference in these two times — sent time and arrival time — contributes to the additional overhead of applying CMM/SS. Note that the propagation delay is assumed to be constant for both the baseline and target profiles.

Figure 5.12 contains a distribution percentile graph of the total runs.

5.7   CMM/SS: A General Solution

The technique of CMM/SS has been successfully applied in the telephony domain to two Internet call control protocols: H.323 and SIP. In both cases, PSTN/IN services were to be accessed from Internet endpoints. A case study of our early work on the application of the nascent ideas of CMM/SS to H.323 is presented in [CHI00]. Subsequently, we have formally specified CMM/SS as a technique and applied it to the SIP signaling protocol [GUR02, GUR03d].

The technique has proved its generality for mapping the Q.1204 BCSM into H.323 and SIP. In both cases, we were able to access PSTN/IN services without changing them at all. Also in both cases, we were able to use the IN call layer we developed with one minor change: the buffer to hold the transaction identifier in the shared-state data structure (see Figure 5.8) had to increase in size when we applied the technique to SIP. SIP transaction identifiers (the Call-ID header) are typically much larger than their H.323 equivalents.

Currently, H.323 and SIP are the preferred protocols for Internet telephony. Both of these protocols can benefit from our technique and access PSTN/IN services. In the future, if other Internet telephony signaling protocols dominate, they should also benefit from the CMM/SS technique in the same manner current protocols have.

5.8 Limitations of CMM/SS

It should be noted that the services we have been able to demonstrate using CMM/SS are related to those executed during call setup and teardown. PSTN/IN services that depend on the media (DTMF, voice recognition, etc.) have not been discussed. To a certain extent, media-based PSTN/IN services are somewhat hard to fit in a peer-to-peer-based system such as SIP, primarily because of the manner in which the telephone network and the Internet behave.

Unlike the PSTN, where each switch handling the call also has access to the bearer channel, in Internet telephony, the equivalent of a switch, a SIP proxy, only has access to the signaling information, not the RTP (or media) session associated with the signaling. Although this is generally a benefit, it proves to be a hindrance for executing services that depend on tones or utterances carried in the media stream. Of course, in such cases, the CMM/SS technique can be applied to a B2BUA controlling a media server instead of a proxy. In this way, the media can be forced to detour through the intermediary, but in doing so, we have broken the end-to-end nature of the Internet. And this becomes a philosophical discussion, not a technical one.

Another limitation that borders on the philosophical is the support of mid-call services. Traditional telephone endpoints were relatively simplistic devices, offering only a 12-digit keypad and a hook-flash capability. Once a session was established, the only way to signal the network for additional services was to depress the hook-flash button or key in a certain sequence of digits. Thus, a user, to receive another call while already talking on the phone, would have to depress the hook-flash button. This resulted in a mid-call signal that allowed the switch to put the first call on hold and the new incoming call to be answered. Once both the sessions were established, the user could toggle back and forth by using the flash-hook.

In direct contrast to the SIP call model, the IN call model has many states that aid in the support for such services through mid-call triggers.

However, it may not be entirely desirable to replicate the PSTN mid-call treatment in Internet telephony. We illustrate with an example: Consider the call-waiting service as it is implemented in the PSTN. Because the PSTN endpoints are hindered by a 12-button interface and one line coming into the handset, they provide stimulus for the incoming call to the user — already in a conversation — through an in-band auditory tone. Now, contrast this with how the similar service may be implemented in Internet telephony: an Internet telephony endpoint is far richer in terms of the user interface than is a PSTN handset. If the user is already in a conversation, an Internet telephony endpoint executing on a desktop (or laptop) computer or a personal digital assistant (PDA) could notify the user of an incoming call by a pop-up window. Should the user need to answer the incoming call, it can place the Internet telephony endpoint on hold by pressing a "Hold" button and simply answer the incoming call by pressing the "Accept" button. In Internet telephony, pressing the "Hold" button does not cause a mid-call trigger as is the case with a PSTN handset when the flash-hook is depressed. Pressing the "Hold" button simply causes the Internet telephony endpoint to stop sending and receiving the media stream.

Another example of a mid-call trigger service in the PSTN is conferencing. Such a service uses a sequence of hook-flashes to add a new party to an existing call. By contrast, conferencing may be implemented in an Internet telephony endpoint simply by pressing the "Hold" button, inviting the new party to the call, and pressing the "Conference" button to get the held party into the call. Unlike the simplistic interface of a PSTN endpoint, an Internet telephony endpoint is far richer in terms of the user interface and far more scalable because an extra call is simply an additional RTP stream emanating from (or destined to) the same IP address.

Note that technically speaking, such legacy PSTN/IN services can be provided through the CMM/SS technique (mid-call triggers do cause a service-state handoff; see Figure 5.9 and Figure 5.10). However, the bigger question is this: Is it worth replicating such services in the Internet? We do not have a definite answer to this question, although we do note that the industry is moving in this direction. Customers with broadband connections who use Internet telephony are still required to connect their PSTN endpoint into an Internet access device. Thus, they still use the existing PSTN endpoint capabilities; i.e., for the call-waiting service on their broadband connection, they still have to revert to using the rather simplistic flash-hook interface of the PSTN endpoint. For these endpoints, CMM/SS allows access to such PSTN/IN services as well.

Another shortcoming of CMM/SS was already mentioned in Section 5.3.4: all CMM/SS mappings will be partially complete because for any two random call models, there will exist minimal semantic loss resulting from the translation of the information elements of one signaling protocol to another.

However, as long as CMM/SS does not overtly constrain the call model in L while providing the service, such a limitation can be acceptable.

A final limitation of CMM/SS is its complexity. Quite rightfully, the question can be asked as to why such a complex model is needed if all that is required are simple lookup-type services? In such cases, appropriate points could be found in the call model of the F domain and a query launched at those points. There is not any need to map the call models of F into those of the L domain. This observation would indeed be true if all that was to be accomplished was to get to a select subset of IN services from Internet endpoints. It will quickly become apparent that as the number of IN services an Internet endpoint wants to access increases, this incremental triggering approach itself becomes complex. Furthermore, all the work that has been vested in the PSTN/IN regarding feature interaction [CAM94] would need to be reproduced in the Internet as well. (Feature interaction is a complex problem in telecommunications software. It stems from the realization that the many services operating simultaneously may interact with each other in several ways, not all of which may be benevolent. Internet telephony may, in fact, exacerbate the feature interaction problem because of the potential of services to reside at the endpoints. The lack of a central controlling authority to arbitrate when an interaction occurs actually makes this a more complex problem in Internet telephony [LEN04, Chapter 10].)

A final reason on why such complexity is required is that PSTN/IN services that go beyond the lookup type may need the flexibility of the IN call model that they were designed for. Thus, CMM/SS, although complex at first sight, is durable in the context of providing a native environment for the IN services to execute.

5.9 Related Work

The Call Model Integration (CMI) Framework [VEM00] aims to access services residing in one network from another, just as our CMM/SS does; however, differences exist between the CMI framework and our approach.

CMI establishes a framework to integrate two call models such that services from either domain are available in the other domain through the framework. The manner in which it does so results in a discrete mapping of each state from one call model to an equivalent state in the other call model. Based on our work in this area, we believe that such a discrete one-to-one mapping is extremely difficult to achieve in practice. [VEM00], recognizing this, instructs that states that do not exhibit a one-to-one mapping be effectively split into substates such that the substates enable a one-to-one mapping. Figure 5.13 demonstrates this splitting. In

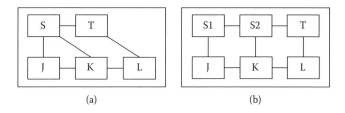

Figure 5.13 Artificial state introduction in CMI.

Figure 5.13a, a one-to-one mapping is not possible between states S and J; portions of S are mapped to K instead. This shortcoming is rectified in Figure 5.13b by dividing S into two substates, S1 and S2, which are then mapped in a discrete fashion.

Introducing an artificial state in this fashion is problematic at best. It raises many additional questions: How does the new state behave in principle with the rest of the states of the call model in which it was introduced? The call model may not be amenable to such an artificial introduction of a new state. How does the designer of the new state decide on the amount of functionality that should be in it, in relation to the state it was carved out from? How easy will it be to realize such a system in working software? Based on such questions, we eschew the approach of introducing artificial states in one call model to make it map to another.

Our approach, by contrast, does not aim to provide a discrete one-to-one mapping between the states of the call models. Because no two call models will be exactly alike, the number of states or transitions will differ between them. Hence, a one-to-many mapping is the best outcome in cases where the cardinality of states differs. Such a one-to-many mapping has an additional burden of subsuming the functionality of many states in one, but this is preferable to the artificial introduction of a new state, which may result in unintended consequences. The mapping shown in Figure 5.9 and Figure 5.10 depicts the one-to-many mapping between states in a SIP state machine and the IN BCSM states.

On a different plane from the call model mapping technique, Miller et al. discuss how to transport TCAP-related signaling in SIP messages [MIL05]. Their work specifies a mechanism by which an eXtensible Markup Language (XML) [W3C04] representation of TCAP messages can be transported in the body of SIP INFO requests [DON00]. This mechanism can be used to allow SIP elements to access features implemented by PSTN equipment without having to implement the binary TCAP protocol. Their work is in contrast to ours, where TCAP messages travel in native form, albeit over the IP network, from the CMM/SS to the SCP. (As an aside,

XML is a meta-language used to describe many different kinds of data. Its primary purpose is to facilitate the sharing of structured data on the Internet. To that extent, XML is self-describing and supports constructs that allow two communicating entities to understand an XML document by validating it against a published schema. Our work described in Chapters 6 and 7 uses XML extensively.)

Literature exists on simple mapping between the states of an Internet telephony call signaling entity and its PSTN equivalent [CAM02, SIN00, VEM02]. However, in all such cases, there does not exist any service state that is shared across the networks. The mappings of [CAM02, SIN00, VEM02] simply depend on discrete messages arriving from one protocol, which are then mapped to an equivalent state of the other one. For example, an incoming SIP INVITE request will be mapped to the other protocol's equivalent signaling primitive to establish a call. There is a complete absence of the notion of service execution in such mappings.

5.10 Conclusion

We have presented a technique to access services in dissimilar networks. The entity making the service request is in a foreign network, in relation to the network that hosts the service (the local network). Thus, the state of a call request with respect to service execution is actually distributed across the two networks. In any distributed system, entity synchronization becomes an important component for the correct and deterministic functioning of such a system. The CMM/SS technique serves to distribute state across the networks and to synchronize the attendant entities as well. The global state of a call is maintained as a composite of each of the individual states. Consistency is imposed by forcing state transitions between the local and foreign networks. Even though the SIP state machine has a smaller number of states and transitions than its PSTN counterpart, this paucity does not translate to an abridged service experience for the user. Using CMM/SS, services written for the PSTN call models can equally well be used with the newer SIP endpoints.

The technique is general enough that in the future, when SIP and H.323 become legacy communication networks, the next generation of signaling protocols should be able to avail themselves of the ideas in CMM/SS to access services from SIP or H.323 networks.

Chapter 6

Crossover Services Originating on the Public Switched Telephone Network

In this chapter, we discuss the second type of crossover service; the events that will lead to the ultimate realization of services of this type occur on the public switched telephone network (PSTN), but the service itself resides in and executes on the Internet.

We have proposed an architecture [GUR04a] to transport discrete events from the PSTN to the user agents on the Internet who have subscribed to such events for service execution. Working closely with the Internet Engineering Task Force (IETF) SPIRITS working group, we have also proposed a set of extensions to Session Initiation Protocol (SIP) [GUR04c, GUR04d] that make it possible to transport discrete events from the PSTN to the Internet. These extensions have been published as an IETF proposed standard [GUR04c]. The architecture and the protocol discussed in this chapter collaboratively provide a common ontology to effectively enable PSTN-originated crossover services.

6.1   Introduction

The Internet has already become a ubiquitous part of our daily life; the telephone has served in that role for an even longer time. Further convergence of these two networks on the services level will lead to innovative service ideas that are not possible in isolation on any one network.

6.1.1   Motivation

The PSTN is a veritable storehouse of events related to users initiating and receiving calls, and cellular phones registering and de-registering their motion across cellular areas. If all these events could be harnessed and transported *out* of the telephone network and *into* the Internet, they could act as catalysts for a wide variety of services.

Consider, for example, presence. Presence can be defined as "the status of devices and applications that create channels for an entity (usually a person) to communicate interactively" [COP01, p. 127]. Presence, as a service, is well defined on the Internet. It is associated with a user (we will call such a user a principal) and is triggered whenever the principal logs into a presence server (like AOL or Yahoo! Messenger). The acts of logging in and logging out indicate the presence and absence, respectively, of a principal. The principal is represented by a Uniform Resource Identifier (URI) (his handle), and the presence or absence state is derived from a device (a computer) associated with the principal.

This kind of interaction, which results in a presence composition, has traditionally been absent on the PSTN. The PSTN can tell if a device assigned to a principal is busy or not, but it cannot leverage this information to compose a presence state associated with a principal. Our work will demonstrate how this is possible by defining an analogous presence service for the PSTN and integrating it with Internet-based presence servers. On the PSTN, the principal would be represented by a phone number (also called a tel URI [SCH04b], which is the standardized format for referring to a PSTN number on the Internet), and the presence or absence of the principal can be derived from his interaction with the device (the actual phone). The PSTN can monitor events occurring on that tel URI to impart the presence state of the principal associated with the tel URI to an Internet-based presence server. If the tel URI corresponds to an office number of the principal, then all of the following acts — lifting the receiver and putting it back on the cradle, making a call, receiving a call — generate events on the PSTN, which can be transported out on the Internet to implement a presence service. The principal is, in essence, present at the office for as long as he is interacting with the office phone.

Now consider availability; it can be defined as a "set of rules and policies, definable by the user (principal), that affect when, how, and by whom contact is made" [COP01, p. 127]. On the Internet, the availability of the principal is typically updated manually. If the principal is, for instance, participating in a telephone conference, he is present but not available to other forms of oral communications. In such cases, he would manually need to set his availability status to "Busy — In a phone call." This example reveals a big disadvantage, which occurs inherently because the principal is interacting with two separate networks: the Internet (for presence and *reflecting* his availability) and the PSTN (which *contributes* to his being available or not). The disadvantage is that the aggregate availability state of the principal cannot be determined by one network alone. Because of the lack of interaction between the PSTN and the Internet, we cannot provide an aggregate availability state of the principal without the principal intervening manually. We term this lack of interaction service isolation. The phenomenon of service isolation is not unique to the wireline network; it is also present in the cellular network.

When a principal turns his 2.5G Internet-capable phone on, it can inform a presence manager, through the Internet connection, to toggle his presence indicator to "on." However, when the same principal initiates (or receives) a phone call, the presence system is unable to reflect his current availability status (i.e., "Busy — In a phone call"). The reason is that the process of initiating (or receiving) a call uses different signaling protocols and a separate voice channel, distinct from the Internet connection. Services using the Internet connection do not interact with the services on the voice channel to provide yet more innovative benefits of integrated networks. Thus, it is impossible to derive a complete state of the principal based on only using one network and its protocols; more intelligence is required.

These examples demonstrate the potential for an architecture that would be general enough to provide this and other more complex crossover services.

6.1.2 Genealogy and Relation to Standards Activities

The idea for PSTN-originated crossover services was first suggested as an outgrowth of a service that actually predates the idea itself. Internet call waiting (ICW), the first attempt at a PSTN-originated crossover service, was prototypical [BRU99a]. In this service, the PSTN kept track of the fact that a principal was utilizing the phone line to get on the Internet. When the PSTN received a call destined to the phone line that was thus busy, it would use the Internet to route a session setup request to the principal's computer. A specialized server running on the computer would cause a pop-up to appear on the screen detailing the name and number of the caller as well as disposition options (see Figure 6.1).

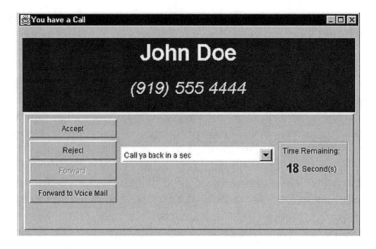

Figure 6.1   ICW screen interface.

The principal could choose to "Accept" the incoming call, thus disrupting the Internet session. In this case, the specialized server would send a message to the PSTN to transfer the call to the principal's line and immediately disconnect the modem connection, thus causing the line to ring. Alternatively, the principal could choose to "Reject" the call or "Forward" it to an alternate number.

In parallel to the implementation of ICW, we foresaw the need for an open interface between the PSTN and Internet to support other novel services [BRU99b]. A preliminary architecture to address this was presented at the 44th IETF [GUR99]. The architecture was further ratified [BRU00] and influenced by the ongoing work in ICW, a key service. However, because many of the protocols that would be used in PSTN-originated crossover services were in mid- to late stages of specification and development, none of the ICW implementations interoperated across vendor boundaries [LUH00]. In 1999, the IETF sanctioned an official working group called SPIRITS [IET05] to inquire into how services supported in the Internet can be started from the PSTN.

We have been active participants in the working group on two levels: first, we have been instrumental in specifying the SPIRITS protocol [GUR04c] as an extension to SIP, and second, we have leveraged our contributions in the working group to further refine and implement our architecture. The working group produced a logical architecture, outlined in [SLU01]. We applied that logical architecture to a physical manifestation first discussed in [GUR99] and have, over the course of our work in this area, refined it to produce the architecture discussed in this chapter.

6.1.3 Contributions

There are two key contributions in this chapter: First, we propose an open architecture built on extensions to standard protocols for PSTN-originated crossover services. The architecture and the associated extensions to the SIP allow us to transport discrete events occurring in the PSTN to the Internet for powerful service execution in the latter domain. The architecture addresses the problem of service isolation, which we outlined previously.

The PSTN-originated crossover service architecture resembles a distributed software architecture, as described in [ROS01]. Such architectures employ distributed middleware (Common Object Request Broker Architecture (CORBA), RMI to design systems. However, we eschew these middleware technologies in favor of standard signaling protocols for call control and data/state transfer. Services are best executed when the service execution platform has unfettered access to the signaling information; application programming interfaces (APIs) tend to shield the programmer from the details of the signaling protocol. Thus, the second contribution of this work is to establish our use of SIP as a distributed middleware component for shared-state Internet telephony services.

The rest of this chapter is organized as follows: The next section outlines our proposed architecture for realizing PSTN-originated crossover services. Realizing such an architecture poses a number of research challenges; we discuss such challenges in Section 6.3. Due to the open environment of the Internet, information exchange requires self-describing data; Section 6.4 proposes a semantic schema for describing the PSTN events. We then specify our proposed extensions to SIP in Section 6.5. Collectively, Sections 6.4 and 6.5 provide a common ontology within our domain. Section 6.6 demonstrates the enabled services through a series of examples. In Section 6.7 we establish a taxonomy for PSTN-originated crossover services. Such taxonomies aid developers in rapid prototyping and refined implementations. In Section 6.8 we present our case for the use of SIP as a distributed middleware in the telecommunications domain. Following that, we take a look at related work in this area and provide conclusions.

6.2 Architecture for PSTN-Originated Crossover Services

PSTN-originated crossover services originate in the PSTN, but at a later time, they cross over into the Internet for subsequent service fulfillment.

In such services, both networks — PSTN and Internet — are involved as follows: An Internet host informs the PSTN that it is interested in the occurrence of certain events, for instance, the event might be an attempt to call a certain PSTN number. When the said event occurs, the PSTN takes a snapshot of the state of the call and transfers this to the Internet host. The latter entity can execute arbitrary services upon the receipt of the notification. Thus, the state of the service is distributed across the two domains and some form of synchronization and a protocol are required to transfer the state of the service from the PSTN to the Internet for execution.

There are three conditions for a service to be considered a PSTN-originated crossover service:

1. *Subscription*: An Internet host subscribes to an event of interest in the PSTN
2. *Action*: The PSTN, during its normal course of operations, under-takes certain actions that lead to the occurrence of the event
3. *Notification*: The PSTN notifies the Internet host of the event and the service itself is executed on the Internet. Depending on the taxonomy of the service, it may be completely executed on the Internet, or the service execution may be shared between the two networks, as was the case with ICW.

A target architecture must thus support Internet hosts subscribing to events of interest occurring in the PSTN and the subsequent notification of the concerned Internet host about the said event of interest by the PSTN.

Given the background, we now propose our architecture for realizing PSTN-originated crossover services that meet the three conditions outlined above. The architecture is deceptively simple, and in keeping with the Internet tradition, it distributes the intelligence to the edges. In fact, the entire PSTN is simply viewed as an Internet user agent (UA) to provide crossover services. Figure 6.2 depicts the architecture.

The architecture is based on separating the network on which the service executes from the one that provides events required for service execution. The service itself is executed entirely on the Internet, but the events that lead to the execution of the service occur on the PSTN. Wireline and cellular telephone networks present a rich palette of events upon which Internet services can be built: registration, mobility, and text messaging are some of the events beyond normal call control that can influence Internet services.

Our architecture, as depicted in Figure 6.2, uses the publish/subscribe mechanism that has proved to be well suited for an event-based mobile communication model [CUG02, MEI02]. User agents (software programs)

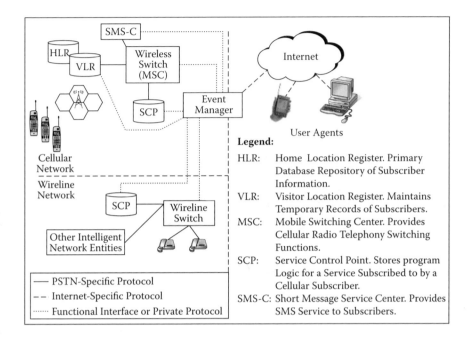

Figure 6.2   PSTN-originated crossover services architecture.

on the Internet subscribe to events on the PSTN. When the event occurs, the PSTN notifies the UA that executes the desired service. The centerpiece of the architecture is the Event Manager (EM), which straddles both networks. It insulates the PSTN entities from Internet protocols and vice versa. It is also responsible for maintaining the subscription state so it can transmit notifications when an event subscribed to transpires.

Figure 6.2 depicts the EM as a stand-alone entity; however, in reality, it may be physically co-resident on the Service Control Point (SCP) or a switch. Our architecture does not limit where the EM is actually located. The only aspect our architecture requires is that the EM has a communication path to the entities in the network that will be generating events. Thus, Figure 6.2 depicts the EM connected to the various entities using dotted lines; the dotted lines represent a functional interface if the EM is co-resident on a certain entity; otherwise, they represent some message passing protocol, the details of which are immaterial to the architecture. The EM should also be able to set dynamic detection points in the SCP (see discussion in Section 3.1.5).

Figure 6.2 shows the PSTN domain on the left-hand side of the diagram and the Internet domain on the right-hand side. The PSTN domain consists of both cellular and wireline networks. Entities on these networks generate events during normal operations; it is these events that need to be captured and transported to the Internet for service execution. The service will execute on the Internet user agents.

Although the architecture appears simple enough, there are research issues that must be addressed. These are cataloged next, along with means to combat them.

6.3   Research Challenges

There are numerous research issues that must be addressed before the architecture of Figure 6.2 can be fully realized. We now enumerate these areas and how they impact our understanding of the problem.

6.3.1   Choosing Target Events

The first challenge is to understand PSTN processing to derive discrete events that can be readily subscribed to using the well-known sub-scribe/notify paradigm. The set of target events thus derived can be har-nessed for crossover services. There are three distinct classes of such events: call-related events, noncall-related events, and application-specific events.

6.3.1.1   Call-Related Events

Call-based events occur in the PSTN as a direct result of making or receiving a call. Anytime a PSTN principal picks up a wireline phone or initiates a cellular session, call-related events occur. For such events, we leverage the PSTN/Intelligent Network (IN) Basic Call State Model (BCSM) we outlined in Chapter 3. As noted in that chapter, the PSTN/IN BCSM is equally applicable to both the wireline and cellular aspects of the PSTN. Thus, we can exploit the rich functionality of the PSTN/IN BCSM to execute crossover services. Each DP in the BCSM becomes an event of interest that can activate a crossover service; Table 6.1 contains a list of all such call-related events. In the table, the first column contains the event name; the second column, a description; and the third column, the DP number relative to Figure 3.6 and Figure 3.7.

6.3.1.2   Noncall-Related Events

Noncall-related events do not require the establishment of a session. Certain events in the cellular network, like cellular phone registration and cellular phone movements, are examples of such events. They do not

Table 6.1 Call-Related Events

Event Name	Description	Figure/DP
OAA	Origination Attempt Authorized: The caller is allowed to initiate a call. Under some conditions (e.g., the use of the line is restricted to a certain time of the day), such a call may not be placed	2.7/DP 3
OCI	Origination Collected Information: The switch has received all the digits from the caller	2.7/DP 5
OAI	Origination Analyzed Information: The switch is attempting to analyze the digits to arrive at the routing information	2.7/DP 7
ORSF	Origination Route Select Failure: The switch could not route the call due to network congestion	2.7/DP 8
OTS	Origination Terminal Seized: The switch has received a message from the terminating side that the called party is being alerted	2.7/DP 14
OA	Origination Answer: The called party has answered the call	2.7/DP 16
ONA	Origination No Answer: The called party did not answer the call	2.7/DP 15
OCPB	Origination Called Party Busy: The called party was contacted but was busy	2.7/DP 13
OMC	Origination Mid Call: Trigger for mid-call services for the caller	2.7/DP 18
OAB	Origination Abandon Call: The caller hung up the phone before the call was completed	2.7/DP 21
OD	Origination Disconnect: The caller disconnected the phone after the call was over	2.7/DP 19
TAA	Termination Attempt Authorized: The terminating switch verifies whether the called party is able to receive this call (i.e., the called party's line has no restrictions against accepting this type of call and the media capabilities are compatible with the caller's)	2.8/DP 24
TFSA	Termination Facility Selected and Available: The terminating switch is attempting to select a resource to reach the called party	2.8/DP 26
TB	Termination Busy: The called party is busy	2.8/DP 25
TA	Termination Answer: The called party answered	2.8/DP 30
TNA	Termination No Answer: The called party did not pick up the phone within a predetermined time	2.8/DP 29
TMC	Termination Mid Call: Trigger for mid-call services for the called party	2.8/DP 32
TAB	Termination Abandon: An erroneous condition occurred while processing the call	2.8/DP 35
TD	Termination Disconnect: The called party disconnected the phone after the call was over	2.8/DP 33

have a counterpart in a wireline network, but this distinction can, in fact, be harnessed to provide powerful crossover services. For example, when a principal turns her cellular phone on, a registration event is generated, which can be propagated to an Internet host for executing presence-based services. Likewise, when a principal enters a predefined geographic zone, a location event is generated that can also be propagated to an Internet host to deliver specific geolocation services. Our proposed architecture is thus transparently able to capture the actions that happen in cellular networks as well and exploit these for subsequent crossover services.

We identify two classes of noncall-related events: registration/de-registration events (to provide presence-based services) and mobility events (for location-based services). For de-registration, we further specify if it occurred due to principal activity (i.e., the principal powered the cellular phone down) or due to network activity (i.e., the network de-registered the principal due to ancillary concerns). Registration always occurs when the cellular phone is turned on. Timer-based or autonomous registration occurs at periodic intervals — ranging from ten minutes to one hour — while the cellular phone is turned on. The granularity of autonomous registrations is typically transmitted to the cellular phone by the serving mobile switching center (MSC) [GAL97, pp. 162–163]. Thus, when the principal moves into a new service area, registrations inform the home network of the current location.

Mobility events are further categorized into two: mobility in the same visitor location register (VLR) area and mobility in a different VLR area. The difference between them is illustrated in Figure 6.3. A VLR area represents the part of the cellular network that is covered by one MSC and VLR combination. Figure 6.3 shows two MSC/VLR service areas. Mobility events associated with principal A occur in the same VLR area, whereas those associated with principal B occur in a different VLR area.

Table 6.2 lists the noncall-related events. The registration-specific events are taken from the Wireless IN (WIN) location registration function state

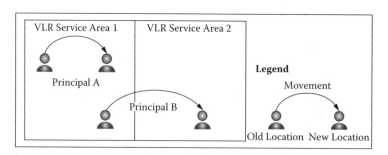

Figure 6.3 Mobility in VLR areas.

Table 6.2 Noncall-Related Events

Event Name	Description	Figure/DP
LUSV	Location update in the same VLR area	N/A
LUDV	Location update in a different VLR area	N/A
REG	Cellular phone registration	2.9/MS_Registered
UNREGMS	Principal-initiated de-registration	2.9/Deregistered
UNREGNTWK	Network-initiated de-registration	N/A

machine we depicted in Figure 2.8. Mobility-specific events do not correspond to a standardized state machine; however, the MSC is informed whenever the location of a mobile host is updated. Thus, as a triggering point, this event can be subscribed to for location-based service execution.

6.3.1.3   Application-Specific Events

The last category is application-specific events. These are, in a sense, the hardest to categorize, primarily because they depend on a specific application, and thus may vary between applications. For instance, the arrival of a Short Message Service (SMS) is an application-specific event that can be leveraged for crossover services; the SMS can be transformed to an instant messaging (IM) and routed out toward the Internet. Similarly, the fact that the remaining balance on a prepaid card is approaching a preset threshold is an application-specific event that can result in a crossover service; an electronic mail or an IM can be sent to the owner of the prepaid card.

Application-specific events are not governed by a call model and its attendant detection points. However, as long as the PSTN is able to detect the event, it should be possible to subscribe to it. We will outline examples in later sections that demonstrate crossover services based on application-specific events.

6.3.2   Modeling PSTN-Originated Crossover Services as a Wide Area Event Notification Service

Our problem space can be characterized by observing that a PSTN-originated crossover service architecture is a system of heterogeneous entities; the entities in the PSTN network generate events, and the entities in the Internet actively seek them out and consume these events. There are two ways of designing such a distributed system: the synchronous pull-based approach and the asynchronous push-based one. Both of these approaches have two main actors: the producer and the consumer. The former produces and advertises the events in the system, and the latter subscribes to these and consumes them.

In the classical pull-based approach, a consumer desiring instantaneous updates to information would need to continuously poll the producer, thus leading to resource contention on both the producer and consumer, network overload, and congestion. This model is adequate for a local area network with a handful of consumers and producers, but it does not scale well to the large networks, like the PSTN or the Internet, and it is not suitable for dynamic (introduction of new event sources) and unreliable environments (loss-prone transports like User Datagram Protocol (UDP)) [CAR01, CUG02, GAD03].

The push-based approach is characterized by the producer proactively notifying the consumers of the event as soon as the event occurs. Such infrastructures are called event notification services [BAC00, ROS97] and are possible alternatives for dealing with large-scale systems [CAR01, GAD03]. In such systems, an additional actor called the broker, or event dispatcher, is involved. The broker is responsible for collecting subscriptions and forwarding notifications to consumers. The architecture we proposed in Figure 6.2 can now be overlaid against the main actors in an event notification service; Figure 6.4 depicts this matching.

The producer of the events includes all the entities in the PSTN — SCP, home location register (HLR), VLR, switches, Short Message Service Center (SMS-C), and others. The consumers of events include the entities on the Internet (the Internet user agents). Producers publish events by sending them to the broker; the EM plays the role of the broker in our architecture. Consumers send an event filter to the EM, which uses this filter to carry out a selection process when the events arrive from the producers. The selection process determines which of the published notifications are of interest to which consumers and delivers notifications only to interested clients.

Modeling PSTN-originated crossover services as a wide area notification service is thus advantageous. Our application space is characterized by asynchrony (consumers do not know when producers will generate events), heterogeneity (consumers and producers are on different networks), and inherent loose coupling — all hallmarks of a wide area network notification service.

6.3.3 Representing the Events

Now that we have the events categorized, we need some manner of representing them in a protocol. In a publish/subscribe system that uses events to communicate, event filters provide a means for consumers to subscribe to the exact set of events they are interested in receiving. Before

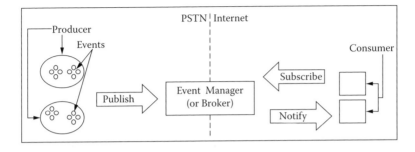

Figure 6.4   Event notification service.

events are propagated, they are matched against the filters and are only delivered to consumers that are interested in them. We represent these event filters as an eXtensible Markup Language (XML) object, which will be encapsulated and transported between the PSTN and the Internet in an appropriate protocol.

To send subscriptions from the Internet host (and notifications from the PSTN) in a standardized manner, we use XML to carry tuples S and N from the Internet to the PSTN, and from the PSTN to the Internet, respectively. An Internet host subscribes to an event of interest represented by a finite tuple $S = (e_v, e_m, e^1_v, e^2_v, \ldots, e^n_v)$, with $n \geq 1$,

where:

e_v = The event that is subscribed to. For events generated as a result of a phone call on the wireline or cellular network, the sets of valid values for e_v are given in Table 6.1. The sets of events in the cellular network not related to a phone call are depicted in Table 6.2.

e_m = The mode of the event; $e_m = \{notify, request\}$. A mode of *notify* requires the PSTN to simply notify an Internet host of the event. A mode of *request* requires that the PSTN temporarily suspend its processing and await instructions from the Internet host on how to further proceed.

$e^1_v, \ldots, e^n_v$ = Additional parameters relevant to e_v. For example, in most cases, one of the parameters sent during subscription will be a phone number for which the Internet host seeks notifications. Any PSTN action that leads to the execution of e_v on that phone number will be of interest to the Internet host.

The notification tuple is represented by $N = (e_v, e^1_v, e^2_v, \ldots, e^n_v)$, with $n \geq 0$. Note that N does not contain the component e_m, and any additional information besides e_v is optional. Table 6.3 lists all the parameters that call-related and noncall-related events can contain. It lists the parameters for a subscription as well as a notification.

Using XML to represent the events pays off when we need to codify and transport application-specific events. Because XML schemas are extensible, application-specific events can be declared in a new namespace and the new namespace imported into the base XML schema dynamically. Obviously, the endpoints employing these extension namespaces will have to agree to the semantics assigned to such events. An XML namespace [W3C99] is a collection of names, identified by a URI reference, which are used in XML documents as element types and attribute names. An XML document may contain elements and attributes that are defined for and used by multiple software modules. Unless appropriate care is exercised, it is highly probable that two software modules define similar elements and attribute names, leading to problems of recognition and collision. The host processing the XML document will be unsure of how to validate such an ambiguous document. XML namespaces alleviate this problem by associating element types and attributes with a universal name.

Table 6.3   Event Parameters

Event	Mandatory Parameter during Subscription	Mandatory Parameter during Notification	Remark
OAA	CallingPartyNumber	CallingPartyNumber CalledPartyNumber	ξ, ψ
OCI	CallingPartyNumber	CallingPartyNumber DialedDigits	○
OAI	CallingPartyNumber	DialedDigits	
ORSF	CallingPartyNumber	CallingPartyNumber CalledPartyNumber	
OTS	CallingPartyNumber	CallingPartyNumber CalledPartyNumber	
OA	CallingPartyNumber	CallingPartyNumber CalledPartyNumber	
ONA	CallingPartyNumber	CallingPartyNumber CalledPartyNumber	
OCPB	CallingPartyNumber	CallingPartyNumber CalledPartyNumber	
OMC	CallingPartyNumber	CallingPartyNumber	
OAB	CallingPartyNumber	CallingPartyNumber	
OD	CallingPartyNumber	CallingPartyNumber CalledPartyNumber	
TAA	CalledPartyNumber	CalledPartyNumber CallingPartyNumber	
TFSA	CalledPartyNumber	CalledPartyNumber	

Table 6.3 Event Parameters (continued)

Event	Mandatory Parameter during Subscription	Mandatory Parameter during Notification	Remark
TB	CalledPartyNumber	CalledPartyNumber CallingPartyNumber Cause	κ
TA	CalledPartyNumber	CalledPartyNumber CallingPartyNumber	
TNA	CalledPartyNumber	CalledPartyNumber CallingPartyNumber	
TMC	CalledPartyNumber	CalledPartyNumber	
TAB	CalledPartyNumber	CalledPartyNumber	
LUSV	CalledPartyNumber	CalledPartyNumber Cell-ID	η
LUDV	CalledPartyNumber	CalledPartyNumber Cell-ID	
REG	CalledPartyNumber	CalledPartyNumber Cell-ID	
UNREGMS	CalledPartyNumber	CalledPartyNumber	
UNREGNTWK	CalledPartyNumber	CalledPartyNumber	

Note:

ξ: CallingPartyNumber is a string used to identify the calling party for the call. The actual length and encoding depend on the dialing plan used; however, it is represented as a string in the XML payload.

ψ: CalledPartyNumber is a string containing the number used to identify the called party. The actual length and encoding depend on the dialing plan used; however, it is represented as a string in the XML payload.

ϒ: DialedDigits contains a nontranslated address (or information) received from the originating user (or line or trunk).

κ: Cause contains a string value of "Busy" or "Unreachable." The difference between these provides services that depend on the called party being busy (engaged) versus unreachable (as it would be if the called party was on the cellular network and the principal was not registered with the network).

η: Cell-ID contains a string used to identify a serving cell identity. The actual length and representation of this parameter depend on the particulars of the cellular provider's network.

6.3.4 Choosing a Protocol

To communicate between the Internet user agents and the EM, we require a protocol that is expressive and extensible, possesses capability description and negotiation primitives, and has transaction-style message exchanges, a flexible naming scheme, and support for event-based communications. In other words, we need a protocol that supports all of the properties listed in Table 4.1.

Protocol expressiveness is a required trait because not all crossover services will result in session setup. Any protocol we choose must be expressive enough to support a wide range of services beyond session startup.

Extensibility is very important to our work. The protocol must be extensible to support arbitrary payload in the signaling messages (the XML object describing subscription and notification filters) and must support asynchronous event notification. The PSTN cannot guarantee when a subscribed to event will occur; thus, the protocol must have primitives to extend subscriptions to pending events or cancel the subscription if it is not needed.

The protocol must also support capability description and negotiation primitives. It must allow the sender of a subscription to describe the payload as well as inform the entity sending out the notification of the capabilities it supports. This allows both entities to communicate in an optimal manner.

A transaction-style message exchange serves to synchronize the entities; thus, this is a desirable property in a protocol. Also important is the support for asynchronous event notifications.

And finally, the protocol must possess a flexible naming scheme. The subscriptions that arrive from the Internet user agents will be destined to a resource in the PSTN and, hence, will contain an appropriate naming scheme (the tel URI [SCH04b], for instance). Notifications, on the other hand, are destined to the Internet user agent, and thus will name a resource on that network using a SIP URI. A protocol that supports tel URIs and other URIs will be extremely attractive.

Based on Table 4.1, the only protocol that supports all our requirements among the three candidate protocols we evaluated is SIP. SIP readily supports arbitrary payload types (it uses Multipurpose Internet Mail Extension (MIME) [FRE96a] to describe the payload) and supports asynchronous event notifications [ROA02]. The events to be subscribed to and the subsequent notification — the tuples S and N — are encapsulated as an XML document. This document is then transported using SIP. A subscription, S, from a UA is encapsulated as an XML object and routed to the EM using SIP. The notification, N, from the EM is also encapsulated as an XML object and routed to the Internet UA over the SIP mesh. Delivering tuples S and N as XML-encapsulated SIP payloads yields a descriptive, extensible, and standards-based codification scheme.

6.3.5 Aggregating Events before Publication

In the wireline network, the source of events is the IN call model executing on the switch. Because there is not any notion of mobility or registration, all wireline events are published from the switch or the SCP connected to the switch. The cellular network is a completely different story, however. Cellular networks have numerous entities that can potentially contribute to event publication. Call-related events are published by the MSC, while an application-specific event such as an SMS queued for later delivery will be published by another specialized server. An important question for an implementation is how to publish the events in a scalable manner.

There are two methods of publishing events: First, each event source acts independently as a publisher and publishes the event toward the consumer. There are two ramifications of this method: (1) the event source must have access to the subscription database and the selection process, and, more importantly, (2) each event source must have a trust relationship with the consumer (Section 6.3.7 covers trust relationships and privacy concerns). The advantage of each event source acting as an independent publisher is the built-in scalability this solution affords.

The second method of publishing events is to have each event source first publish the event to an aggregate point, which, in turn, publishes it to the eventual consumer. The event source and the aggregate point are assumed to belong to the same autonomous organization. The aggregate point collects all events published and runs the selection process on them to determine if a consumer should be notified. Because the consumer is always communicating with the aggregate point for all notifications, this method does not suffer from the problems associated with trust and privacy.

In the architecture of Figure 6.2, the event aggregation point is the EM. Having each event source publish events independently to the consumer leads to a complex system with the same logic replicated in multiple event sources. It is far better to aggregate the events at a centralized location and send notifications out from there.

6.3.6 Scalability of the EM

It is a complex task to gather events in the network. The EM has to react with a number of entities that are generating events, as discussed above. Scalability is a concern if not handled appropriately. We provide a performance study of the EM in Chapter 7, where we discuss the internals of an EM that we constructed. To preview this issue, however, scalability concerns dictate that there is at least one EM for every switch in the system. In other words, the EM must not be shared with more than one switch, and it should be co-located with a switch for maximum performance.

6.3.7 Privacy, Security, and Trust

The events subscribed to and the subsequent notifications may contain extremely private information. The notifications have the potential to reveal sensitive location information or other damaging information (for example, an SMS message from a broker to a client containing an account number). Privacy of this information in transit is of paramount importance.

Besides privacy, another axis of interest is trust: the EM must be sure that subscriptions are coming from an authenticated UA. Transitively, the UA must ascertain that the notifications are coming from an authenticated EM instead of a malicious hijacker acting as an EM.

To authenticate and encrypt communications between two previously unknown parties on the Internet, public key cryptography is the best option. Two known problems with it are key distribution and the lack of a well-known and universally trusted certificate authority (CA). In Chapter 7 we outline a method that mitigates both of these to implement a secure framework using public key cryptography.

6.4 An XML Schema to Represent Events in the PSTN

Peers exchanging information in the open environment of the Internet require the data to be self-describing. In Appendix A, we present an XML schema that can be used to encode the PSTN events in a self-describing and extensible manner. The events of Table 6.1 and Table 6.2 are part of the schema.

The work described in this chapter and our efforts in the IETF SPIRITS working group progressed in parallel to a certain extent; hence, we have chosen to reuse the IETF terminology instead of defining an alternative terminology. Thus, we refer to the schema of Appendix A as a SPIRITS schema and a document validated by it as a SPIRITS XML document. Likewise, when we discuss the SIP extensions, they will be characterized by tokens with a *spirits* prefix, and XML namespaces will contain *spirits* as a component.

The SPIRITS schema supports other namespace extensions, thus allowing application-specific events to be dynamically understood. A detailed look at the elements and attributes of a SPIRITS XML document follows:

6.4.1 The <spirits-event> Element

The root of the XML document is the <spirits-event> element. This element must contain a namespace declaration (xmlns) to indicate the namespace on which the XML document is based. XML documents compliant to the schema we propose must contain the Uniform Resource Name (URN) [BER98] urn:ietf:params:xml:ns:spirits-1.0 in the namespace declaration.

Other namespaces may be specified as needed. We have registered this namespace and the schema itself with Internet Assigned Numbers Authority (IANA) through our work in [GUR04c].

As an aside, a URN is a subset of a URI that is required to remain globally unique and persistent even when the resource it names ceases to exist or become available. A book number assigned by the U.S. Library of Congress is a URN, as is the legal name of an individual.

The <spirits-event> element must contain at least one <Event> element.

6.4.2 The <Event> Element

The <Event> element contains three attributes, two of which are mandatory. The first mandatory attribute is a type attribute whose value is either INDPs or user-prof. These types correspond, respectively, to call-related events and noncall-related events.

The second mandatory attribute is a name attribute. Values for this attribute are limited to the event names defined in Table 6.1 and Table 6.2.

The third attribute, which is optional, is a mode attribute. The value of mode is either N or R, corresponding respectively to (N)otification or (R)equest. The difference between them is the semantics of the service offered. In Notification-style services, call processing continues normally once the notification has been sent out. In Request-style services, call processing is temporarily halted in the PSTN until further instructions are received from the Internet host. That is why synchronization of the attendant entities is an important trait we were looking for in a protocol. The default value of this attribute is N.

If the type attribute of the <Event> element is INDPs, then it must contain at least one or more of the following elements (unknown elements may be ignored): <CallingPartyNumber>, <CalledPartyNumber>, <DialedDigits>, or <Cause>. These elements were defined in Table 6.3 as event parameters. They must not contain any attributes and must not be used further as parent elements. These elements contain a string value.

If the type attribute of the <Event> element is user-prof, then it must contain a <CalledPartyNumber> element and it may contain a <Cell-ID> element. None of these elements contain any attributes, and neither must be used further as a parent element. These elements contain a string value. All other elements may be ignored if not understood.

A SPIRITS XML document will look like the example shown in Figure 6.5. Figure 6.6 dissects the document in more detail. Such an XML document will be present in the subscription as well as the notification SIP signaling messages.

```
<?xml version="1.0" encoding="UTF-8"?>
  <spirits-event xmlns="urn:ietf:params:xml:ns:spirits-1.0">
    <Event type="INDPs" name="OD" mode="N">
      <CallingPartyNumber>5551212</CallingPartyNumber>
    </Event>
    <Event type="INDPs" name="OAB" mode="N">
      <CallingPartyNumber>5551212</CallingPartyNumber>
    </Event>
  </spirits-event>
```

Figure 6.5   XML document corresponding to schema.

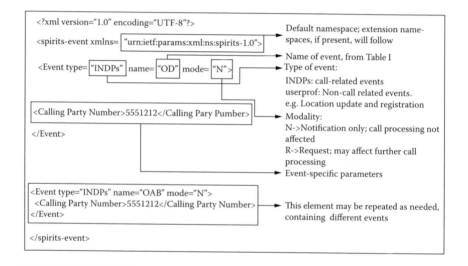

Figure 6.6   Understanding the XML document.

6.5   Proposed Extensions to SIP

We have extended SIP across two axes: first, we specify two SIP *event packages*, and second, we introduce a new MIME type used to describe a payload transported by SIP. Before delving into the details of these extensions, we first present an overview of how SIP handles asynchronous event notifications.

6.5.1   The Asynchronous Event Notification Framework in SIP

The asynchronous event notification framework in SIP is defined in RFC3265 [ROA02] as "provid[ing] an extensible framework by which SIP nodes can request notifications from remote nodes indicating that certain events have occurred."

The RFC3265 framework can be thought of as an abstract base class that defines the overall behavior of the entities, but leaves specific behavior to the classes derived from the abstract base class. What this essentially implies is that RFC3265 provides broad guidelines on what is expected of a SIP entity participating in the framework; the details of exactly how an entity meets those expectations are left to the specific instance derived from RFC3265. These specific instances are called event packages by RFC3265. Thus, RFC3265 simply mentions that consumers must issue a subscription, but does not mandate the payload format of that subscription. Likewise, it requires producers to issue a notification, but again, it does not specify the contents of the payload. Specifying the format of the payload and how it is to be interpreted is performed by the consumers and producers. Out of necessity, they have to agree to a standard format for representing the payload. There are other such instances where RFC3265 provides a general behavior that is further refined by the specific event package; these include details on how long subscriptions last, when they should be refreshed, the rate of notifications, and whether forking (replication of a request across multiple search branches) is permitted.

Figure 6.7 contains an overall message flow that shows how asynchronous event notification works in SIP. An Internet UA, the consumer, sends a SUBSCRIBE request to the EM. The payload of the SUBSCRIBE request is composed of a SPIRITS XML document that contains the filter used during the selection process. The EM accepts the subscription and installs the filter. The protocol requires a final response (200 OK) to be sent to the consumer. At some later time, one or more events will occur that will match the filter using the selection process. At that time, the EM will issue a NOTIFY request that contains another SPIRITS XML document. This document lists all the events that the selection process indicated matched the filter.

Figure 6.7 labels the EM as a producer. Strictly speaking, the EM is not a producer, but rather an aggregate point where all the event sources in the PSTN publish their individual events to. However, logically, it helps to label the EM as a producer because, as far as the consumer (Internet UA) is concerned, the events are published to it by the EM.

6.5.2 The Extensions

We have defined two SIP event packages: spirits-INDPs and spirits-user-prof. The former package corresponds to all events associated with originating/receiving a call, while the second event package corresponds to noncall events (registration, de-registration, location updates, etc.).

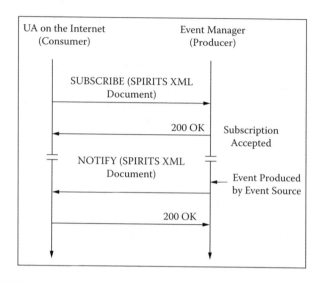

Figure 6.7 **Asynchronous event notification in SIP.**

Both the event packages carry PSTN event-related information in SIP signaling. To foster widespread interoperability, we have also registered a new MIME type called application/spirits-event+xml [GUR04c] and have registered it with IANA. This MIME type defines the body of the event packages. The Content-Type header in SIP SUBSCRIBE and NOTIFY requests will contain this MIME type, and the body will contain a SPIRITS XML document.

6.5.2.1 The spirits-INDPs Event Package

Each event package is given a name that is carried in a special header in SIP called Event header. Call-related events listed in Table 6.1 are named by the token spirits-INDPs.

When a consumer wishes to install a filter for a selected set of events, it forms a SUBSCRIBE request and sends it to the EM. The subscribe request will contain a SPIRITS XML payload. The XML payload may contain one or multiple events; the names of the events are drawn from Table 6.1. Any mandatory parameters for that event specified in Table 6.3 must also be included in the XML payload.

The subscription installs a filter at the EM. The subscription remains fresh for as long as the time period negotiated between the EM and the consumer, or until an event is published that satisfies the selection criteria of that filter. In other words, if the subscription filter contained the events $\{e_1, e_2, e_3\}$ and event e_2 happened to satisfy the selection process, the entire subscription expires. The EM will not wait for events e_1 and e_3 to occur before considering the subscription stale. In such a case, the consumer can always send out a new subscription if it wants to resubscribe to the same events or a subset of the previously subscribed to events. RFC3265 also allows a subscription to be refreshed before it becomes stale. This is done by resending the subscription with the same event filter before the previous subscription has had a chance to expire.

When an event is published that satisfies the selection process at the EM, the EM will send a NOTIFY request to the consumer. The NOTIFY request will contain the SPIRITS XML document. Once a notification is sent out, the subscription expires implicitly.

Notice that the rate of notifications going out of the system is fairly constant for each consumer. The average number of calls that a principal makes (or receives) per hour is small enough that network or resource congestion for that principal is not an issue. As an aggregate, the number of notifications going out will be a function of the number of consumers who have subscribed to these events. We will present detailed performance studies of the system in Chapter 7.

6.5.2.2 The spirits-user-prof Event Package

Noncall-related events listed in Table 6.2 are named by the token spirits-user-prof. As was the case with call-related events, consumers sent a filter to the EM containing the events of interest. The SPIRITS XML document may contain one or multiple events; the names of the events are drawn from Table 6.2. Any mandatory parameters for that event specified in Table 6.3 must also be included in the XML payload.

When an event is published that satisfies the selection process at the EM, the EM will send a NOTIFY request to the consumer. In the NOTIFY request will be an XML payload corresponding to the schema outlined in this chapter. Unlike the spirits-INDPs package, once a notification is sent out, the subscription does not expire. To understand why, consider mobility as an event in the filter. If the principal and his cellular phone generating the mobility event happen to be in a high-velocity car, the system will have to send a massive amount of notifications in a short period. If the subscription was to be expired by the first such notification, the consumer would be forced to reinstall the subscription and involve the system in another round of filter installation overhead. This, of course, leads to a verbose protocol and irresponsible use of network resources. To avoid such scenarios, we propose a ceiling on the number of notifications that should be transmitted under the spirits-user-prof package.

Figure 6.8 contains an algorithm that throttles the rate of notifications if the published rate of the events *per principal* exceeds one event per

```
offset ← 15; // seconds

Process_Events
    (principal, event_list) ← get next event; // Blocks
    if (selection_process(principal, event_list)) then
        if (Should_send(principal))
        then
                send_notification(principal, event_list);
        else
                discard event_list;
        end if;
    end if;

Should_send(principal)
    cur_time ← get current time;
    last_sent ← get_last_notification_time(principal);
    if (last_sent + offset <= cur_time) then
        return 0;
    else
        last_sent ← cur_time;
        set_last_notification_time(principal, cur_time);
        return 1;
    end if;
```

Figure 6.8 Throttling algorithm.

15 seconds. Producers must send a notification of a given type toward the same principal only once every 15 seconds. We will revisit this issue from a performance perspective in Chapter 7.

In this package, subscriptions do not expire on the publication of an event that satisfies the selection process; thus, it would appear that once installed, a subscription always remains active. However, this is not the case. Unless they are refreshed, subscriptions become stale and expire automatically after the time duration negotiated between the consumer and the EM has passed.

6.6 Examples

We now present some examples showing how the proposed SIP extensions and the architecture function in operation. All of the services involve one or more principals; A is a principal on the Internet, and B and C are principals on the PSTN. A may be a person, in which case he executes a user agent that implements the protocol described in this chapter. A

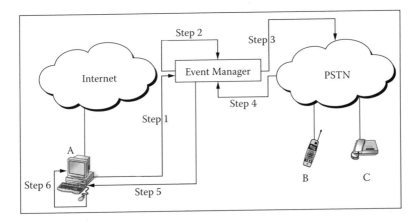

Figure 6.9 Operational view.

may also be an automaton, in which case the automaton itself is a user agent that implements extensions to the protocol that we have described. Under certain service scenarios, A and B may refer to the same principal.

Figure 6.9 contains a step-by-step view of the operations we describe next. The steps in the figure correspond to the steps listed in the following discussion. The user agent, when executed on the Internet, sends a subscription to the EM containing a list of desired events it wishes to receive a notification for (step 1); in essence, the user agent is sending an event filter to the EM. The EM creates a subscription based on the event filter, stores the event filter in a database for the selection process to be executed later, and interfaces with the PSTN entities (step 2) such that when the event is generated in the PSTN, it is published to the EM (step 3). As events occur on the PSTN, they are published to the EM for mediation (step 4).

The EM runs the selection process on the events to determine which ones will result in a notification. Matching events result in a notification sent out of the EM (step 5). The user agent, in turn, executes a specific service applicable to the event notification (step 6).

In the examples below, readers are urged to study the event filter in the body of the SIP message and correlate the mandatory parameters of individual events as per Table 6.3.

6.6.1 *Notification of Missed Calls*

IM is a service that is not generally associated with the wireline PSTN, although the cellular PSTN has supported a similar service in the form of SMS for some time. (To be pedantic, differences exist between IM and

SMS. For one, SMS messages are limited to 160 to 200 characters, whereas IM systems in deployment today are capable of carrying larger messages. Furthermore, the network stores an SMS for later delivery if the recipient is not able to get the message in real time. IM systems, on the other hand, vary in capabilities from discarding the message if the recipient is not present to queuing it in a relay for later delivery.)

Instant messages have been used in one form or another as long as the Internet has been around. In the early stages of Internet, the UNIX *write(1)* command caused a text message to be sent from the sending terminal to the recipient terminal, where it would show up instantly on the screen. More sophisticated uses of IM technology developed with the advent of the Internet in the enterprise and home markets. However, traditionally this service has not been associated with the PSTN. We now show how it becomes a crossover service when applied to the PSTN.

In this service scenario, A wants to receive notifications of calls destined to her PSTN desk phone. Presumably, A is going to be in a meeting and cannot receive phone calls to her desk, but would like to know who called her. She runs an Internet UA that sends a SUBSCRIBE request, portions of which are reproduced in Figure 6.10.

Of importance in the figure are two SIP headers, Content-Type and Event, and the payload. The Event header contains a token referring to one of the extensions we proposed, spirits-INDPs. The Content-Type header contains the MIME type that binds the document to be a SPIRITS XML document. The document identifies one event, TAA. This event, which is a detection point in the Terminating BCSM (T_BCSM) half of the PSTN call model, is published when an incoming call arrives at a certain phone line (6305550216) in the PSTN. Upon publication of this event to the EM, the EM will send out a notification to A, portions of which are reproduced in Figure 6.11.

```
SUBSCRIBE sips:6305551212@provider.com SIP/2.0
...
Event spirits-INDPs
Content-Type: application/spirits-event+xml
Content-Length:...

<?xml version="1.0" encoding="UTF-8"?>
  <spirits-event xmlns="urn:ietf:params:xml:ns:spirits-1.0">
    <Event type="INDPs" name="TAA" mode="N">
      <CalledPartyNumber>6305550216</CalledPartyNumber>
    </Event>
  </spirits-event>
```

Figure 6.10 Subscription for missed calls.

```
NOTIFY sips:vkg@iit.edu SIP/2.0
...
Event: spirits-INDPs
Content-Type: application/spirits-event+xml
Content-Length:...

<?xml version="1.0" encoding="UTF-8"?>
<spirits-event xmlns="urn:ietf:params:xml:ns:spirits-1.0">
    <Event type="INDPs" name="TAA" mode="N">
       <CalledPartyNumber>6305550216</CalledPartyNumber>
       <CallingPartyNumber>8475551212</CallingPartyNumber>
    </Event>
</spirits-event>
```

Figure 6.11   Notification of missed calls.

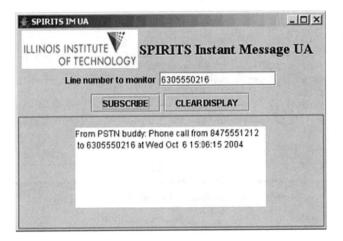

Figure 6.12   Graphical user interface for a missed-call notification.

When A's user agent receives this notification, it can inform A through an audiovisual reminder by beeping and presenting a pop-up window with the pertinent information. Figure 6.12 contains an example of such a graphical user interface.

6.6.2 Presence for a Principal Using a Wireline PSTN Endpoint

In this service scenario, A is interested in receiving the presence information of a principal B through B's interaction with the wireline PSTN phone. As an attribute, presence for a phone line representing B can be deduced if B answers a phone call, makes a phone call, or simply lifts the phone from its cradle and puts it back. The interaction of B with the phone device to derive a presence service can be summed up by subscribing to the following two events: OAA and TA.

The first event, which is drawn from the Originating BCSM (O_BCSM) half of the PSTN call model, represents B as the party that initiated the call, and the second event, drawn from the T_BCSM half, represents B as the party that received the call. The act of lifting the phone to make a call will result in the publication of event OAA. Similarly, if B receives a call and answers it, TA will be published. Accordingly, the payload contains the XML event filter in Figure 6.13.

```
SUBSCRIBE sips:6305551212@provider.com SIP/2.0
...
Event: spirits-INDPs
Content-Type: application/spirits-event+xml
Content-Length:...

<?xml version="1.0" encoding="UTF-8"?>
<spirits-event xmlns="urn:ietf:params:xml:ns:spirits-1.0">
    <Event type="INDPs" name="OAA" mode="N">
      <CallingPartyNumber>6305550216</CallingPartyNumber>
    </Event>
    <Event type="INDPs" name="TA" mode="N">
      <CalledPartyNumber>6305550216</CalledPartyNumber>
    </Event>
</spirits-event>
```

Figure 6.13 Subscription for wireline presence.

```
NOTIFY sips:vkg@iit.edu SIP/2.0
...
Event spirits-INDPs
Content-Type: application/spirits-event+xml
Content-Length:...

<?xml version="1.0" encoding="UTF=8"?>
<spirits-event xmlns="urn:ietf:params:xml:ns:spirits-1.0">
    <Event type="INDPs" name="OAA" mode="N">
      <CallingPartyNumber>6305550216</CallingPartyNumber>
      <CalledPartyNumber>8475551212</CalledPartyNumber>
    </Event>
</spirits-event>
```

Figure 6.14 Notification for wireline presence.

When one of OAA or TA is published to the EM, the EM sends out a notification to A's user agent. A's user agent can then toggle the presence state of the B in a graphical user interface of some kind. We will revisit the implementation of this service in more detail in Chapter 7. For the sake of completeness, Figure 6.14 contains the notification sent out assuming that the event OAA got published to the EM.

6.6.3 Presence for a Principal Using a Cellular PSTN Endpoint

In this service scenario, A is interested in receiving the presence information of a principal B through B's interaction with the cellular phone. Compared to determining the presence of principals on the wireline PSTN, doing so for those on the cellular network is much simpler. A only needs to subscribe to one event: REG. This event will be published by the cellular PSTN network as soon as B powers on his cellular phone and it registers with the network.

Figure 6.15 contains the subscription sent out by A. There is one point worth studying: note that the Event header contains spirits-user-prof, a token that refers to the second of our proposed extensions to SIP. Subscriptions of this have different requirements on expiry and staleness; they do not become stale after the event is published. They will, of course, expire normally after their expiration time has been exceeded.

When the REG event is published to the EM, a notification is sent out to A's user agent. Figure 6.16 contains portions of such a notification. It is instructive to note that there is an additional parameter — Cell-ID — present in the notification document (as per Table 6.3). The Cell-ID

```
SUBSCRIBE sips:6305551212@provider.com SIP/2.0
...
Event: spirits-user-prof
Content-Type: application/spirits-event+xml
Content-Length:...

<?xml version="1.0" encoding="UTF-8"?>
<spirits-event xmlns="urn:ietf:params:xml:ns:spirits-1.0">
    <Event type="userprof" name="REG" mode="N">
       <CalledPartyNumber>6305550216</CalledPartyNumber>
    </Event>
</spirits-event>
```

Figure 6.15 Subscription for cellular presence.

```
NOTIFY sips:vkg@iit.edu SIP/2.0
...
Event: spirits-user-prof
Content-Type: application/spirits-event+xml
Content-Length:...

<?xml version="1.0" encoding="UTF-8"?>
<spirits-event xmlns="urn:ietf:params:xml:ns:spirits-1.0">
    <Event type="userprof" name="REG" mode="N">
      <CalledPartyNumber>6305550216</CalledPartyNumber>
      <Cell-ID>182A9</Cell-ID>
    </Event>
</spirits-event>
```

Figure 6.16   Notification of cellular presence.

provides an aspect of geolocation that can be used for location-based services; we will revisit this in Chapter 7 with a concrete example.

6.6.4   Helping First Responders

Often, in cases of extreme emergencies (power blackouts, tornadoes, and hurricanes), the PSTN experiences abnormal call loads that it was not designed to handle. In such situations, the caller typically hears a "fast busy" signal, which signifies that the PSTN is unable to route the call further because all trunks emanating from the central office (CO) are busy. This condition, epitomized by the ORSF event, can be used to propagate critical information to first responders.

In such a scenario, a first responder configures a list of peer first responders that includes alternative methods to reach the peers. This may include electronic mail or an IM. The PSTN saves this list on an automaton (we will call it A). A is preconfigured to subscribe to the ORSF event. During the emergency, when the first responder attempts to place a call through the now congested telephone network to other first responders, A will receive a notification from the EM. Upon receipt of such a notification, A can proactively send out electronic mail or IM messages to peer first responders allowing them to know that an emergency may be in process.

Note that such a system can also be used to inform family members that their loved one is safe. In such a scenario, the list stored on A will contain alternative means to reach family members. When a user, in an emergency, attempts to call a family member but gets a busy signal, A can step in and inform the family member of the safety of their loved one.

6.6.5 Schema Extension: Notifications for Low Prepaid Card Balance

Our final example involves extending the XML schema. In this service scenario, a principal B has a prepaid card with a PSTN provider. B would like to be notified when the prepaid balance falls to within 20 percent of the limit. Furthermore, B would like to be notified through an electronic mail message. To that extent, an automaton, A, on the Internet subscribes to a low prepaid card balance event. Figure 6.17 contains a subscription filter that addresses these constraints.

The event filter in Figure 6.17 contains an extension namespace (whose alias is ppb) with its own set of elements and attributes. In the event filter, there are four elements in the ppb namespace: ppb:Event, ppb:number, ppb:limit, and ppb:remind. The ppb:Event element refers to the class of events that the namespace may support; in the example, the name of the specific event is "prepaid." ppb:number refers to the phone number of the primary prepaid card holder, ppb:limit refers to the low watermark after which a reminder is sent to the URI specified in ppb:remind. The URI in the example is a "mailto" URI, which will result in A sending an electronic mail to the primary prepaid card holder informing him that the card is within 20 percent of being depleted.

6.7 A Taxonomy of PSTN-Originated Crossover Services

To impose some organization on PSTN-originated crossover services as well as help implementers in characterizing such services for rapid implementation,

```
SUBSCRIBE sips:6305550216@provider.com SIP/2.0
...
Content-Type: application/spirits-event+xml
Content-Length:...

<?xml version="1.0" encoding="UTF-8"?>
<spirits-event xmlns="urn:ietf:params:xml:ns:spirits-1.0"
               xmlns:ppb="http://www.provider.com">
    <ppb:Event name="pre-paid">
       <ppb:number>6305550216</ppb:number>
        <ppb:limit>20</ppb:limit>
        <ppb:remind>mailto:vkg@iit.edu</ppb:reminder>
    </ppb:Event>
</spirits-event>
```

Figure 6.17 Subscription for low prepaid card balance.

we attempt to taxonomize PSTN-originated crossover services. By and large, the taxonomy is suggested by the e_m element of tuple S. In other words, PSTN-originated crossover services can be categorized in two classes: *notification* and *dialog oriented*. The latter automatically implies the former; the reverse is not true.

Notification services are the simpler of the two. The PSTN simply notifies the Internet host of the occurrence of the event of interest. Once the notification is sent out, call processing continues normally in the PSTN without further aid of the Internet host. It should be pointed out that the notification may not be the result of call processing at all. For instance, in cellular networks, a notification may be sent to an Internet host when a principal turns on his or her cell phone, thus registering with the network, or a principal on the wireline phone network may simply lift and set down the receiver on the cradle. A notification can be sent to the Internet host that executes an appropriate service, such as toggling the presence and availability state of a principal on the cellular or wireline telephone network.

Dialog-oriented services are executed when the Internet host receives an INVITE request from the PSTN (the ICW service discussed in Section 6.1.2 is a good example). The Internet host acts as an extended SCP to the switch, as the latter has temporarily suspended call processing until it gets further instructions from the Internet host. Services under this classification scheme may exhibit long delays and mandate strict timing behavior on the part of the Internet host. If the Internet host expects a fair amount of time (in the order of seconds) to generate further instructions, it should periodically send messages (provisional responses in SIP) to the switch to reset any relevant timers in the PSTN. The PSTN, on the other hand, should start tearing the call down and reclaiming resources if it does not get any response from an Internet host (the Internet host may have crashed, or it could be misbehaving).

Dialog-oriented services can be further subclassified as follows:

> **Static dialog**: Under this classification, the Internet host establishes a relationship with the switch, thereby effectively controlling the switch until the service is executed. Using ICW as an example again, call processing is suspended at the switch until the Internet host makes a final determination on the disposition of the call. This disposition is sent to the PSTN in the form of a final response to the INVITE request. There are two distinguishing facets for this classification: First, the DPs are armed *a priori* on the switch; in other words, a SUBSCRIBE may not be needed (ICW implementation experience [LUH00] confirms this). The second factor is that once the Internet host has sent a final disposition, the relationship between the switch and the Internet host effectively terminates.

Dynamic dialog: The key property of this classification is that the Internet host maintains an ongoing relationship with the switch even after sending a final disposition to the INVITE request. It can, for instance, choose to get subsequent events from the PSTN by arming successive DPs after a call has been established. For example, the Internet host may subsequently subscribe to the hang-up event, i.e., have the PSTN notify it when the call is terminated. A static dialog may transition to a dynamic one based on the service aspects of the Internet host.

We believe that this taxonomy will aid in better understanding PSTN-originated crossover services and that the classification outlined here helps form a standard reference template for implementation design issues.

6.8 SIP: The Distributed Middleware

The term *middleware* refers to the software layer between the operating system and the distributed applications that interact via a network. By this definition, middleware includes the actual communication protocol used by the peers, reliability of the protocol, security of the protocol, and scalability.

The Internet's open ecosystem places further burdens on the middleware operating within its environs [GEI01]:

1. The interacting peers belong to independent, autonomous organizations that do not necessarily trust each other.
2. Communications between peers take place over an insecure medium.
3. The communication infrastructure does not provide quality of service guarantees. Messages between too many peers over Transmission Control Protocol (TCP) may impose unreasonable overhead, whereas the unreliability of UDP may not be desirable.
4. Because of working in an open environment with a multiplicity of peer implementations, exchanging information requires self-describing data and agreement on common ontologies.
5. The peers may frequently be mobile, thus changing their network identifiers when needed.

For existing middleware solutions, this list of properties poses new demands.

Invocation-based middleware systems like CORBA or Java remote procedure call (RPC) are extremely useful for building distributed systems; however, as Gaddah and Kunz [GAD03] discuss, although these models are adequate for a local area network, they do not scale up to the Internet. The model of such systems is built around a reactive pattern: an object remains passive until an operation is performed on it. This type of a model only supports one-to-one correspondence of the peers and involves a tight coupling between the peers because of the synchronous nature of the communication pattern. Because our problem space is characterized by the asynchronous push-based approach (see Section 6.3.2), we do not consider invocation-based middleware systems to be a good solution to our problem.

The category of event-based middleware [BAC00] is especially applicable to our problem domain. Such middleware systems address the requirements of decoupled, asynchronous interactions in large-scale, widely distributed systems [BAC00, CAR01]. Event notification is the basic paradigm used by such middleware. Events contain data that describes a request or a message, and are generated at an event source and propagated from notifiers, which may be the same event source or a separate entity, to subscribers. In fact, we have already described our architecture in these terms in the discussion of Section 6.3.2.

There are several examples of event-based middleware: CEA [BAC00], Siena [CAR01], JEDI [CUG01], and ToPSS [CUG02]. We researched them to determine if they are applicable to our problem domain by applying them to a set of requirements pertinent to our domain. These requirements constitute the qualities we were searching for in a middleware solution: First, authentication of the communicating peers and encryption of the traffic were mandatory. Second, it was desirable to have a solution that allowed the events to be transported in an alternative protocol than the one dictated by the middleware. For instance, in our domain, all consumers are SIP enabled; they use the protocol from initiating sessions to executing services. If the event-based middleware systems supported the transport of events within an alternative protocol (say, SIP), we would not have to load and execute another piece of software on our endpoints to send out event subscriptions and receive event notifications. In certain cases — limited-memory PDAs — loading and executing new software may not be possible, and in other cases, it may not be desirable to add more communication complexity to the mix. And finally, reliability and fault tolerance of event notification service should be addressed. The SIP endpoints may be mobile and constantly attaching and detaching themselves from network access points to acquire new network identifiers. The event notification protocol should not fail in the face of such uncertainty.

Two major components of event-based middleware are the expressiveness of the event-matching kernel and the speed at which events can be processed. All the middleware systems we reviewed excelled at both of these. The area where they fell short was security. None of the middleware we researched accounted for the chaotic nature of the Internet, where new identities can be obtained many times over to allow a questionable cloud of repudiation to hang over the communicating parties.

The middleware framework that provided a security solution closest to our requirements was CEA. However, CEA addressed security through an adjunct architecture called Oasis; the security infrastructure was not integrated into the event-based middleware itself. Oasis indeed uses certificate-based credentials; the certificate is tied to a specific service. To use a service, the user must present the certificate to the server. However, the manner by which the user receives the certificate, i.e., how the user herself is authenticated, is not well specified. Furthermore, the certificates Oasis uses are different than the ones used by Internet protocols like Hypertext Transfer Protocol (HTTP) and SIP. These protocols use X.509 certificate-based [ITU00b] authentication and encryption schemes through Transport Layer Security (TLS) [DIE99] to secure sessions and authenticate the endpoints.

Another disadvantage of these frameworks when used in our domain was that the protocol used to transport the event subscription and notification between the peers was distinct from SIP. In certain cases — like Siena — the intercommunication protocol can be HTTP or SMTP, and with some work, we could have modified Siena to use SIP as well. But Siena did not match up against other requirements, namely, security and reliability.

We were unable to use any of the existing event-based middleware systems. The disadvantage this presented was that it precluded us from leveraging the extensive event-matching kernels of such systems. But on the other hand, we felt that a general solution for our problem domain should be structured around SIP because the protocol provided us with all the tools we needed: we could transport events in a secure and encrypted manner by using TLS in SIP. We could also use its advantage of running over multiple transports: TCP and UDP (the protocol provides an application layer retransmission mechanism to guarantee delivery if UDP is being used). An active research issue in building a middleware consists of developing a robust notification protocol that supports guaranteed delivery of messages despite transport considerations (too many notifications over TCP impose unreasonable overhead, whereas the unreliability of UDP may not be tolerable) and network failures [CUG02]. Our use of SIP mitigates this issue.

Additionally, we could use fault tolerance and a redundancy scheme built into the protocol, which depends on Domain Name Service (DNS) SRV resource records [GUL00]. SRV resource records are special DNS records that allow administrators to use several servers for a single domain (for load balancing and fault tolerance), to dynamically move services from host to host (for reliability), and to designate some hosts as primary servers for a service and others as backups. And finally, we could use the asynchronous event notification extension built into SIP to transport discrete events and, at the receiving endpoints, extract these events and handle them appropriately. (That is, when a consumer sends the events, the producer saves them for using as a filter during the selection process, and when a producer sends the events, the consumer executes a service made possible by the event.)

For all the reasons presented, we developed a SIP-based middleware framework, which we will discuss in Chapter 7.

6.9 Related Work

The closest effort related to our work is PSTN to Internet Interworking (PINT) [PET00]. PINT describes an architecture and protocol that is a mirror image of our work. Whereas our work aims to transport discrete events from the PSTN to the Internet for service execution on the Internet, PINT transports service requests from the Internet to the PSTN for service execution in the PSTN. Thus, clicking a link on a Web page (Internet) would cause a service request to travel to the PSTN, which would set up a call between two parties. This allowed such services as the following:

> Click-to-Dial: While browsing through a company's Web site, clicking on a Web link causes the PSTN to make a call between the Web user and a customer service representative of the company. Request-to-Fax: Clicking on a Web link causes the PSTN to send a fax to a certain destination. As an example, a restaurant's Web site may contain a link that, when pressed, transmits a facsimile of the menu.
> Request-to-Hear-Content: Clicking on a Web link causes the PSTN to call a certain number and arrange for some content to be spoken out.

Berkeley's ICEBERG project [WAN00] integrates telephony and data services spanning diverse access networks. Their approach is expansive because their architecture maintains an overlay network consisting of different geographic ICEBERG points of presence (iPOP) and many ICEBERG access points to isolate the access network from the overlay network. The iPOPs are coordinated by a centralized clearinghouse that serves as a bandwidth broker and accountant (loosely akin to our EM, although unlike ICEBERG, the EM does not perform bandwidth brokering). Our approach, by contrast, is extremely lightweight and follows the service mantra of the Internet, whereby the core network is used simply as transport and services are provided at the edges. In a sense, the entire PSTN is abstracted as a UA generating and sending events to another UA that then executes the services.

Weinstein [WEI02] and Buddhikot et al. [BUD03] outline how wireless local area networks (W-LANs) complement rather than compete with cellular mobile systems. However, they view this work from a data perspective, i.e., providing data connections in the cellular mobile system with a guaranteed quality of service. Buddhikot et al.'s work clusters around allowing an IEEE 802.11 hotspot operator to access the profiles and policies of a 3G user when the latter roams into the former's network. They do not discuss the architecture we have proposed here to transport discrete events from one network to the other for service execution.

6.10 Conclusion

We have implemented PSTN-originated crossover services for the wireline [GUR03a] and cellular [GUR04d, GUR05a] components of the PSTN. Our implementation will be discussed in detail in the next chapter, where we will highlight its applicability to ongoing research in the area of pervasive computing.

It is important to note that the ontology we have described is not limited to PSTN events culled from the BCSM. The methodology presented here is independent of any call model; just an agreement is needed to specify the points where a call model is amenable to interruption. Once this is done, an extension schema can be constructed and the framework we discussed here used to transport the events between networks. For example, Kozik et al. [KOZ03] use our proposed architecture and protocol extensions to transport encoded Parlay events between the PSTN and the Internet.

A key requirement of PSTN-originated crossover services will be third-party programmability of such services. Arguably, the service creation framework for the World Wide Web (WWW) infrastructure has thrived because it enables third parties to provide value-added services over a common transport, namely, Internet Protocol (IP). The most important factor for the success of WWW services has been a common lingua franca (HTTP/Hypertext Markup Language (HTML)) and an extensive service creation tool set (Web Common Gateway Interface (CGI), Active Server Pages, Java scripts, servlets, Service Object Access Protocol (SOAP),[1] etc.). Telephony, on the other hand, has traditionally been an environment where

---

[1] SOAP [W3C03] is a lightweight protocol intended for exchanging structured information in a decentralized and distributed environment. It uses XML to represent a message construct that can be exchanged over a variety of underlying protocols.

the inner workings of the protocols and services, although not entirely secret, were not subject to as much public access and scrutiny as Internet protocols have been. We believe that the Web model of allowing open, well-defined protocols needs to be replicated for PSTN-originated crossover services. To that extent, the work presented in this chapter contributes to an open and extensible architecture for crossover services based on standard protocols to help third parties in developing such services.

We believe that establishing a taxonomy of PSTN-originated crossover services is extremely important, so that implementers can quickly identify various techniques for rapid implementation. Thus, we have proposed a taxonomy of PSTN-originated services.

And finally, we presented a case for using SIP as a distributed middleware.

Chapter 7

Smart Spaces in the Telecommunications Domain

By the year 2025 the entire world will be encased in a communications skin, according to experts at Bell Laboratories. "We are already building the first layer of a mega network that will cover the entire planet like a skin," declared Arun Netravali, then president of Bell Laboratories in 1999 [LUC99]. "As communication continues to become faster, smaller, cheaper and smarter in the next millennium, this skin, fed by a constant stream of information, will grow larger and more useful."

The merging of the Internet and the public switched telephone network (PSTN) is one facet contributing to the realization of that communication skin. In this chapter, we leverage our architecture and the protocol extensions of Chapter 6 to create a pervasive computing infrastructure in the telecommunications domain. There are many definitions of pervasive computing, ranging from enabling embedded devices for effective inter-communication to the more abstract definition of allowing a user to make a knowledgeable decision based on the quick, efficient, and effortless flow of information between communication and computing entities.

7.1 Introduction

Pervasive computing has been elegantly defined as the "creation of environments saturated with computing and communication yet gracefully integrated with human users" [SAT02, p. 10]. To date, a case could be made that the most successful technology to be integrated with the daily life of a human user has been the telecommunication infrastructure, consisting of both general and special-purpose computers providing cellular and wireline PSTN access. Of late, another form of communication has been added to this mix: the Internet.

The emergence of the Internet and its application to existing PSTN infrastructure opens up previously unexplored avenues of research into pervasive computing. Consider, for example, the following scenarios: Bob works for an advertising company in Chicago and keeps in touch with his colleagues at other locations through the cellular Short Message Service (SMS). This morning, Bob did not bring his cell phone to work, so he will miss all SMS messages destined to his cell phone. When Bob arrived to work in the morning, he discovered that he has to attend an all-day meeting scheduled at the last minute. He calls his wife to tell her about his plans only to find that his home phone is busy. Furthermore, Bob is expecting an important call from his brother, which he will likely miss while he is in the meeting. And finally, Bob's manager is flying in to meet him and Bob would like to be notified of her presence as soon as the manager arrives at Chicago airport and turns on her cellular phone. Bob would also like to track the location of the manager as she makes her way to Bob's office. This will allow Bob to better prepare for the meeting by taking advantage of any extra time afforded to him if the manager gets delayed by traffic.

Given that Bob's productivity is tied to a device — a PSTN cellular or wireline phone — the absence of the device hampers him. However, Bob has access to another ubiquitous network, the Internet. It would be extremely helpful if both networks, the PSTN and the Internet, cooperated to benefit Bob in the following manner: when Bob receives an SMS for his cell phone that is not in his possession, the PSTN network intercepts the SMS and transforms it to an instant message (IM) to be delivered to another of Bob's devices connected to the Internet. This could be a laptop, desktop, or wireless Institute of Electrical and Electronics Engineers (IEEE) 802.11-capable personal digital assistant (PDA). The fact that Bob's home phone is busy can be leveraged by the PSTN to provide real-time updates of his phone line and inform him through a buddy system application,[1] so Bob knows when his wife has finished her conversation — something

---

[1] Similar to Yahoo! Messenger, AOL Buddy List, or Microsoft Messenger, except that the PSTN phone line is used as a primary device to deduce the presence and availability of a person using that phone line.

akin to a presence and availability indicator for his home phone line. Likewise, when his brother calls Bob's cellular or office phone, the PSTN can inform him through an unobtrusive IM on his PDA. And finally, when Bob's manager arrives in Chicago and turns on her cellular phone, the PSTN can intercept this event and send a discreet IM to Bob (or toggle the presence status of the cellular device associated with the manager in Bob's buddy list application). As the manager travels from the airport to the office, the cellular network can inform Bob of her progress so he can adjust his schedule accordingly if she gets caught in rush-hour traffic.

These examples demonstrate the potential of services that leverage both communication networks. In isolation, instant messaging and completing a phone call are simply atomic services, but when combined in this manner, their utility increases many times more than if they were simply operating alone.

If the intent of pervasive computing is indeed, as Mark Weiser predicted [WEI91], to make computers an invisible part of our daily lives, providing us constant information in unobtrusive ways to help us reach informed decisions, then we are that much closer to his vision now than we have ever been. The availability of wireless communication technologies (e.g., the cellular phone network, IEEE 802.11, Bluetooth) and portable communication endpoints (e.g., wireless phones and PDAs) and the emergence of the Internet to augment the PSTN have opened up unprecedented avenues of research into how to harness these technologies for a seamless communications experience.

Satyanarayanan [SAT01] makes an observation on pervasive computing that rings true for our scenario as well: perhaps the biggest surprise in the scenario we presented previously is how simple and basic all the component technologies are. The hardware technologies — wireless phones and PDAs, cellular phones (and software underpinnings), IM servers, presence servers, and location servers — all exist today. However, why do we not see such services in use today? The real answer, Satyanarayanan theorizes, is that the whole is much greater than the sum of its parts; the real research is in the *seamless integration of component technologies* (author's emphasis). In this chapter, we provide this seamless integration to make the services we outlined possible.

We implement the architecture we proposed in Chapter 6 that allows services of the kind we detail above. We attempt to create, in effect, a telecommunications *smart space*. A smart space [SAT01] is an aggregate environment composed of two or more previously disjoint domains. In a smart space, one domain senses and controls the other. Our work demonstrates one domain (the Internet) controlling events occurring in another domain (the PSTN) to execute services. We were motivated by the simple fact that users utilize both networks on a constant basis, yet these networks do not interact with each other to provide ubiquitous services of the kind outlined in the opening scenario. Thus far, the PSTN has been mainly used to access the Internet (through modem pools), and the Internet to digitize and transport voice packets. We feel that both networks can benefit from a cross-pollination of ideas at the services level. In our work, we embed pervasive communications infrastructure in both of these disjoint spaces to allow one to sense and control the other.

The rest of this chapter is organized as follows: In Section 7.2, we outline the research thrusts that are important to the agenda of pervasive computing and communications. We then apply the architecture and protocol extensions discussed in Chapter 6 to create a telecommunication smart space. We will introduce the main actors, the relationship between them, and a privacy model that allows secure communication between the participating entities. Section 7.4 presents a detailed look into an important component of our work: the Event Manager (EM), discussed in Chapter 6. We outline its design and implementation. We then discuss the performance of the architecture by analyzing the behavior of the EM. Section 7.6 discusses related work followed by a conclusion.

7.2 Research Thrusts of Pervasive Computing

Satyanarayanan [SAT01] offers four research thrusts in pervasive computing. We discuss these next and will revisit them at the end of the chapter to observe how close our architecture adheres to these principles.

7.2.1 Effective Use of Smart Spaces

A smart space is an aggregate environment created from two previously disjoint spaces. A space may be an enclosed area (meeting room, corridor) or an open area (courtyard, park). Embedding computing infrastructure in building infrastructure allows one world to sense and control the other. The quintessential example of this is automatic temperature adjustment in a room based on the occupant's stored electronic profile. Smartness may also extend to individual objects, whether located in a smart space or not.

7.2.2 Invisibility

The second research thrust is invisibility of the pervasive computing technology from the user's consciousness. Although the ideal is complete disappearance of such technology, in reality, the best approximation is minimal user distraction.

7.2.3 Localized Scalability

Scalability is a critical problem in pervasive computing. Depending on the specifics of the smart space implemented, the intensity of interaction among cooperating entities may increase. Although this may be acceptable for a smart space that is confined to a small area, it is prohibitive for one that spans geographical distances. The problem is further compounded if one or more of the cooperating entities are mobile and thus may be limited by the bandwidth, energy, and computing power.

In pervasive computing, the density of interaction has to fall off as the distances between the cooperating entities increase. Good systems design has to achieve scalability by severely reducing interactions between distant entities.

7.2.4 Masking Uneven Conditions

The rate of penetration of pervasive computing technology will vary considerably. The capabilities of entities that provide services to users invisibly will vary considerably. One way to reduce the amount of variation observed by users is to have their computing space compensate for "dumb" environments — in essence, provide a canonical representation of their computing space and ensure that a service, if it is operating in a nonfriendly environment, at least tunes its behavior to fit the circumstances.

In applying these four precepts to our work, we consider the telephone network and the Internet to be two disjoint worlds. Although the telephone network has been used to connect users to the Internet (through modems), and the Internet has been used to transport digitized voice, these networks have not cooperated at the services layer to a great extent. We will demonstrate how our architecture allows these disjoint worlds to come together and enable the sensing and control of one world by another, almost invisibly, by using localized scalability and masking uneven conditioning.

7.3 Implementing a Telecommunications Smart Space

The components of an architecture required to implement a telecommunications smart space were established in Chapter 6. To reiterate, the architecture itself was presented in Figure 6.2; crucial decisions, including choosing the target events in the PSTN, representing them through XML, the choice of a protocol to transport these events between the networks, and the protocol extensions to do so, were all discussed in detail in

Chapter 6. In this chapter, we fill in the missing pieces as we realize the architecture of Figure 6.2 in the context of pervasive communications.

7.3.1 The Main Actors

With a protocol chosen and an architecture defined, we now present a usage model to help understand the various players involved in the execution of service in our smart space.

There are four parties of interest in a smart-space service: the PSTN service provider, the Internet service provider, the end user of the service (to minimize overloading the term *end user*, we will call such an end user a *consumer*), and the principal (recall, a principal was defined in Chapter 6 as the user on the PSTN whose device — a phone — generated the events a consumer would be interested in).

The PSTN service provider owns and operates the PSTN network on which events are generated. The events are generated by a device associated with a principal. The consumer is the party in the Internet Protocol (IP) domain that requests the PSTN service provider to send it events of interest for service execution. Finally, the Internet service provider is the party that provides the IP transport to the consumer. The PSTN service provider and Internet service provider can belong to the same organization, but they do not have to. As a general rule, we will assume that they are not part of the same organization.

To use a service in our smart space, we envision that a specialized user agent (UA) will be made available to consumers by the PSTN service provider or a third party working with the service provider. The specialized UA, in addition to supporting the base Session Initiation Protocol (SIP) functionality [ROS02], will also support our proposed extensions to SIP, detailed in Section 6.5 and [GUR04c]. The specialized UA will be preconfigured with the address of an EM in the domain of the PSTN service provider that will be contacted for all services. Furthermore, it is not expected that the consumer will be conversant with XML to formulate event of interest or interpret the notification. Rather, the PSTN service provider will codify the events it supports in a graphical user interface (GUI) to make it easier for the consumer to choose events of interest. The specialized UA will construct the appropriate XML document based on the selection and send it to the EM at the preconfigured address.

7.3.2 Authentication and Encryption

The events contained in subscriptions and subsequent notifications consist of extremely private information. The notifications have the potential to reveal sensitive location information or other damaging information; e.g., an SMS message from a broker to a client may contain an account number, a mobility event may contain location-specific information that can be misused, or seemingly innocuous events such as a user dialing a certain number at a certain time of the day may have privacy implications if this information ends up in the wrong hands.

Privacy of information in transit is paramount. Another axis of interest here is trust: the EM must be sure that subscriptions are coming from an authenticated UA. Transitively, the UA must ascertain that the notifications are coming from an authenticated EM instead of a malicious hijacker acting as an EM.

To authenticate and encrypt communications between two previously unknown parties on the Internet, public key cryptography is the best option. Public key cryptography, sometimes called asymmetric encryption, was invented in 1976 by Whitfield Diffie and Martin Hellman. It is a cryptographic system that uses a pair of mathematically related keys — a public key and a private key. The public key is widely disseminated, while the private key is jealously guarded. These keys have the property that the private key decrypts only what the public key encrypts. Furthermore, knowing the public key in no way compromises the integrity of the private key; i.e., the private key cannot be guessed from the public key. Besides encryption, public key cryptography can be used to create digital signatures, which authoritatively identify the parties. Further information on cryptography in general, including public key cryptography, can be found in [STA02].

Two known and interrelated problems with a global public key infrastructure are key distribution and the lack of a well-known and universally trusted certificate authority (CA) [GUT02, GUT04]. A CA is usually an organization that, for a fee, will issue the key pair and vouch for the authenticity of the public key; i.e., the CA attests that the public key contained in a certificate actually belongs to the owner specified in the certificate. Key distribution is the act of disseminating as widely as possible the public keys. There are many approaches for doing so: public keys can be exchanged through electronic mail, hosted on personal Web sites, or submitted to centralized key distribution centers, like a CA.

The weakest link is in the trust chain. Should the CA err in vouching for the identity of a user, a security breach would occur because a malicious user who used a forged identity has successfully obtained a valid certificate from a CA and can do untold damage. Imagine a company that uses a forged identity to obtain a certificate. Now imagine an unsuspecting user who conducts business with the company and, in good faith, sends them his credit card number and other vital information in an electronic mail encrypted with the public key of the company. The public key infrastructure has been used to infringe on the privacy of the unsuspecting user.

Continuing with issues in the trust chain, in large-scale deployments, user Alice may not be familiar with user Bob's CA, so Bob's certificate may include his public key signed by a higher-level CA, which is presumably recognizable by Alice. Thus, public key infrastructure implicitly assumes a hierarchy of CAs. Currently, there does not exist one global CA on the Internet; many companies perform the role of a CA, and it is very likely that keys issued by two different CAs would need the arbitration of a higher-level CA.

Regardless of the problems, public key cryptography is the most secure and scalable solution to encrypt communications on an open network and vouch for identities between parties who may be completely unknown to each other. We present a solution that mitigates the two problems of public key cryptography that we identified at the onset — lack of a central CA and key distribution.

We assume the worst-case scenario: all parties are unknown to each other. The consumer authenticates herself to the PSTN service provider using a credit card, driver's license, or preexisting customer relationship with the service provider. Using the identity provided by the consumer, the PSTN service provider's service management system creates two keys — P_{pr}(Consumer) and P_{pu}(Consumer) — corresponding to the private and public keys, respectively (Figure 7.1a). The management system then burns the public key of the PSTN service provider (P_{pu}(PSP)) and P_{pr}(Consumer) in the UA; P_{pu}(Consumer) is also escrowed at the PSTN service provider. The UA arrives to the consumer through a download link or mailed on a disk.

When the consumer subscribes to certain events, the transmission stream is encrypted using P_{pu}(PSP); subsequently, it is decrypted using P_{pr}(PSP), which is already in the possession of the PSTN service provider. When the service provider sends a notification, P_{pu}(Consumer) is used to encrypt the contents and P_{pr}(Consumer) is used to decrypt them (see Figure 7.1b). Recall that P_{pr}(Consumer) was configured earlier in the UA. In this manner, message integrity is maintained. To prove the identities, both the PSTN service provider and the UA can digitally sign the message using their private keys. Note that if a well-known CA is not available, the PSTN service provider can act as a CA and self-sign the certificates it issues.

The technique we present here has broader implications. The lack of a centralized CA trusted unequivocally by all participants has meant complexity in providing a reasonable certificate distribution infrastructure [BUR04, GUT02, GUT04]. One manner by which to mitigate this problem is to use an established central authority who has a preexisting relationship with the users for whom keys need to be issued. A central authority that

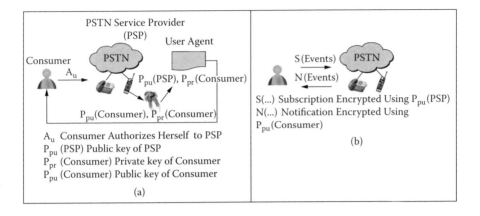

Figure 7.1 Authentication and encryption process.

fits this description is the PSTN operator. Such an authority already has an established relationship with the customers it services, and thus establishing the identity of the customer beyond a reasonable doubt is much easier. Conversely, the customers are apt to trust a certificate issued by the PSTN operators rather than by a CA they may not be familiar with.

We believe that this model of leveraging an existing central authority that is trusted by the users diminishes some of the problems traditionally associated with public key infrastructures.

7.3.3 Policies

The previous section dealt with ensuring the integrity and confidentiality of the data in transit and authenticating the endpoints. There is one final step before the entire system can be rendered usable: policies of the principal.

A policy, in our smart space, can be defined as a directive that allows consumers to access the events of a principal. For example: Allow consumer Vijay K. Gurbani access to my location between 5:00 P.M. and 7:00 P.M. every day. The principal can set such policies through a Web interface or by calling the PSTN service provider's customer service center. Because a relationship already exists between the PSTN service provider and the principal, it can be used to authenticate the principal to the service provider and install specific policies.

Policies are represented as tuples in our system and stored in a persistent store that the EM has access to. When an event is published, the EM runs the selection process on the published event and the policies to determine a match. The elements in our policy tuple, P_{α}, include

$$P_\alpha = \{C_\alpha,\ C\text{-}URI_\alpha,\ C\text{-}Ev_\alpha,\ Pr_\alpha,\ Ev_\alpha,\ B_\alpha,\ E_\alpha,\ D_\alpha,\ M_\alpha,\ W_\alpha\}$$

where:

C_α = Consumer's identity

$C\text{-}URI_\alpha$ = Consumer's URI — the URI to which notifications are sent

$C\text{-}Ev_\alpha$ = List of events (comma separated) that a consumer is interested in getting a notification for

Pr_α = Event source (principal's device)

Ev_α = List of events (comma separated) that the principal grants a consumer

B_α = Begin time (military style)

E_α = End time (military style)

D_α = Day within a month (1 to 31)

M_α = Month of the year (1 to 12)

W_α = Day of the week (1 to 7, 1 = Monday)

When a policy is initially created, it will contain a value for C_α; however, $C\text{-}URI_\alpha$ and $C\text{-}Ev_\alpha$ elements will be null (because a consumer has not yet subscribed to the set of events in Ev_α). When the consumer subscribes to the set of events, $C\text{-}URI_\alpha$ will be populated with a URI that the consumer can be reached at and $C\text{-}Ev_\alpha$ will be populated with a list of events that the consumer is interested in. The consumer's identification was provided by the principal when the policy was first created. For a subscription to be accepted by the system, this identification has to agree with the identification stored in the consumer's certificate.

The last three elements in the tuple are to be interpreted as the corresponding fields in the UNIX *crontab* file; i.e., they can be a single number, a pair of numbers separated by a dash (to represent a range of numbers), a comma-separated list of numbers and ranges, or an asterisk (a wildcard that represents all valid values for that field). The elements B_α and E_α could also contain an asterisk that signifies that Ev_α is valid for a particular C all 24 hours.

As an example, the verbal policy of a principal, expressed as "Allow consumer Vijay K. Gurbani access to my location between 5:00 P.M. and 7:00 P.M. every day," is translated into the following tuple:

```
P_a = {"Vijay K. Gurbani",
       null, null,
       6305551212,
       "LUDV, LUSV",
       1700, 1900,
       *, *, *}
```

Note that the event list, Ev_α, consists of two events — LUDV and LUSV — which are associated with mobility in the cellular network (see Section 6.3.1.2). The consumer's identity is provided; thus, when the consumer subscribes to these events, his authenticity will be questioned before the subscription is accepted. Because a consumer has not yet subscribed to these events, the second and third elements of the tuple P_α are not specified. The start and stop times are self-explanatory. Because the principal specified "everyday" in the verbal policy, the elements D_α, M_α, and W_α of the tuple are set to their wildcard values. The event source itself (principal's device) is specified as the second element in the tuple.

Once a descriptive policy has been reduced to its corresponding tuple, the policy tuple is saved in the persistent store, where it will be updated by a consumer subscription and subsequently used during the selection process to match an event.

7.3.4   Constructing a Telecommunications Smart Space

Recall that in a smart space, one world senses and controls the other one. The UA running on the Internet interfaces with the PSTN to subscribe to a set of events that it is interested in. The EM saves this subscription in persistent store. When such an event is published in the PSTN, the EM runs the selection process using the published event and the policies stored for the consumer. If the selection process results in a match, the consumer's UA is notified. The UA thus senses and controls the events in the PSTN. We now present a set of five services that demonstrate the telecommunications smart space. The overall flow of the service is similar to that depicted in Figure 6.9.

The five services demonstrate the example scenario we opened the chapter with — the interaction of the consumer, Bob, with the events in the PSTN of various principals that he is interested in. The examples below employ many phone numbers that represent the event sources in the PSTN. Table 7.1 correlates a phone number (the event source) to the principal it represents.

In our implementation, encryption and authentication were provided by OpenSSL v0.9.7b. OpenSSL is an open-source library (freely available at http://www.openssl.org) that implements the Transport Layer Security (TLS) specification [DIE99]. Using OpenSSL, a secure socket is created between two communicating entities such that all information flowing through that socket is encrypted and the identities of the communicating endpoints authenticated. We acted as a self-signed CA and issued certificates for the PSTN service provider and all the user agents in the system.

Table 7.1   Correlation of an Event Source to a Principal

Event Source	Principal
718-555-1212	Bob's manager
425-555-1212	Bob's brother
815-555-1212	Bob's home line
630-555-1212	Bob's cellular phone
847-555-1212	Bob's desk phone

Policy tuples were stored in a SQlite v3.0.7 database. We loaded the database with 1 million (1M) policy tuples. SQlite is a public-domain small C library (< 250 KByte code space) that implements a self-contained, easy-to-embed-in-applications, zero-configuration Structure Query Language (SQL) database engine. It is freely available at http://www.sqlite.org.

Our laboratory consisted of wireline switches and phones connected to them, other wireless switches, simulated cells, base stations, and cellular endpoints that received signals from the base stations. Using simulation tools, we are able to simulate the signal attenuation that reproduces the movement of cellular subscribers between cells. Figure 7.2 depicts the laboratory setup.

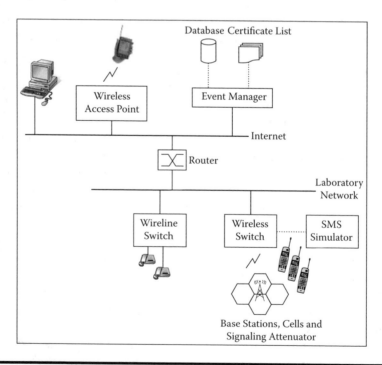

Figure 7.2   Laboratory setup.

There were a number of wireline phones connected to the wireline switch. Each such phone published events when the principal it belonged to interacted with it. These events were published to the EM, which analyzed them for a pending subscription. Likewise, the wireless switch had a number of wireless phones connected to it through a bank of equipment that simulated base stations and cells. A principal interacting with the cellular phone would publish events that made their way to the EM for mediation. The bank of equipment that simulated cells and base stations also came with software that allowed us to attenuate the signal strength monitored by the cellular phones. We could thus simulate movement of the principal by calming the signal strength in one cell and intensifying it in the adjoining cell. The cellular phone would sense the intensity of the new signal and use it instead. We also had access to an SMS simulator, which would send an SMS message destined to a particular phone to the wireless switch. The recipient phone would be turned off, causing the wireless switch's attempt to page it to fail (in cellular networks, the wireless switch pages the cellular phones in an area to locate them). The failure event, in association with sending an SMS, was trapped and published to the EM. The net result of this would be to simulate a principal's cellular phone being turned off and thus incapable of receiving SMS messages.

In our prototyping, we detected and published all events from the switches for both the wireline and wireless PSTN. This was done out of necessity because it was not always possible to get a configured Service Control Point (SCP), home location register (HLR), or visitor location register (VLR). However, this step in no way compromised the integrity of our architecture or prototype because the EM really does not care from where the event was published. The events published from the switches to the EM arrived over a Transmission Control Protocol (TCP) connection.

7.3.4.1 The Presence Service

In Chapter 6, we argued that the presence service, popular on the Internet, is just as applicable to the PSTN. Whereas on the Internet the presence service is triggered by the principal using a device — a computer — to actively log into the presence system (Yahoo! Messenger, for instance), on the PSTN presence can be deduced from the interaction of the principal with another device — the phone.

Our opening scenario in this chapter demonstrated the need for such a service for our consumer, Bob.

> Bob's manager is flying in to meet him and Bob would like to be notified of her presence as soon as the manager arrives at Chicago airport and turns on her cellular phone.

```
SUBSCRIBE sips:bob@pstn-provider.com SIP/2.0
...
Event: spirits-user-prof
Allow-Events: spirits-user-prof, spirits-INDPs
Accept: application/spirits-event+xml, application/pidf+xml
Content-Type: application/spirits-event+xml
Content-Length:...

<?xml version="1.0" encoding="UTF-8"?>
<spirits-event xmlns="urn:ietf:params:xml:ns:spirits-1.0">
    <Event type""userprof" name="REG">
      <CalledPartyNumber>7185551212</CalledPartyNumber>
    </Event>
</spirits-event>
```

Figure 7.3   Presence subscription for principal.

Accordingly, Bob runs his UA, which sends the subscription shown in Figure 7.3 to the EM.

There are a few items of interest in Figure 7.3. First, Bob's UA is informing the EM of all the event types it can handle through the Allow-Events header (spirits-user-prof and spirits-INDPs, which correspond to the extensions we propose in Chapter 6). This subscription itself corresponds to the spirits-user-prof event package, which transports non-call-related events between the networks.

The second item of interest is the Accept header. The Accept header in Internet protocols such as Hypertext Transfer Protocol (HTTP) and SIP serves to inform the recipient of the Multipurpose Internet Mail Extension (MIME) types the sender can accept and interpret. In Figure 7.3, Bob's UA is able to interpret two MIME types: application/spirits-event+xml and application/pidf+xml. The former MIME type corresponds to SPIRITS XML documents, and the latter corresponds to XML documents that transport presence-related information. The Presence Information Document Format (PIDF) [SUG04] is an IETF standard that describes the format of an XML document conveying the presence state of a principal. We will provide more information on a PIDF XML document later in this section.

The final item of interest is the Content-Type header; this header contains the MIME type — application/spirits-event+xml — corresponding to a SPIRITS XML document. In the XML document, Bob's UA is issuing a subscribe for the REG (registration) event of the principal's cellular phone, identified by the number 7185551212.

Assuming the subscription sent by Bob's UA was accepted by the EM, Bob's UA will update its visual interface (see Figure 7.4). Notice the last row — 7185551212 — is set to "Unavailable;" i.e., the system does not have

Figure 7.4   Depicting presence.

any information about the principal at this time. At a certain point in time after the subscription has been accepted, and while it is still fresh, the principal (Bob's manager) lands at O'Hare Airport. She turns on her cellular phone, which registers with the cellular network and sets in motion further processing of Bob's pending subscription. As soon as the cellular phone registered, the cellular switch published that event to the EM. The EM executed the selection process to determine if any consumer was subscribed to events published by the principal. The selection process returned true for Bob's pending subscription. Subsequently, the EM collects all the relevant information from the database regarding where to send the notification (Bob's UA had imparted this information, which was saved in the database during the subscription phase), creates a notification, and sends it. Figure 7.5 depicts such a notification.

As before, there are a number of items of interest in the notification. First, note the SPIRITS XML document in the payload. This document contains the event that was published (REG) along with the Cell-ID parameter, which is mandatory in the notification according to the rules specified in Table 6.3.

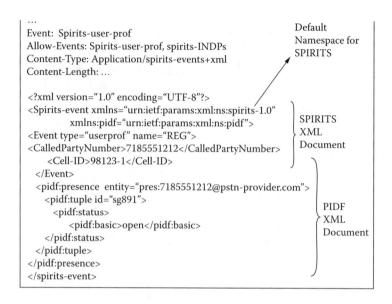

```
...
Event: Spirits-user-prof                                Default
Allow-Events: Spirits-user-prof, spirits-INDPs          Namespace for
Content-Type: Application/spirits-events+xml            SPIRITS
Content-Length: ...

<?xml version="1.0" encoding="UTF-8"?>
<Spirits-event xmlns="urn:ietf:params:xml:ns:spirits-1.0"
        xmlns:pidf="urn:ietf:params:xml:ns:pidf">      SPIRITS
<Event type="userprof" name="REG">                      XML
<CalledPartyNumber>7185551212</CalledPartyNumber>       Document
    <Cell-ID>98123-1</Cell-ID>
  </Event>
    <pidf:presence entity="pres:7185551212@pstn-provider.com">
      <pidf:tuple id="sg891">
        <pidf:status>                                   PIDF
          <pidf:basic>open</pidf:basic>                 XML
        </pidf:status>                                  Document
      </pidf:tuple>
    </pidf:presence>
  </spirits-event>
```

Figure 7.5   Notification containing multiple XML documents.

The second item of interest is the MIME-type negotiation occurring between the consumer (Bob's UA) and the notifier (EM). In Figure 7.3, the consumer indicated the capability to support an additional MIME type, namely, the one corresponding to PIDF documents (application/pidf+xml). Thus, when the notification is sent, a portion of the payload contains a PIDF XML document. A PIDF document contains a number of attributes; we provide enough information to interpret Figure 7.5. Interested readers are urged to consult [SUG04] for an in-depth treatment of PIDF.

A PIDF document is published on behalf of a URI identified in the "entity" attribute of the <pidf:presence> element. In our example, this is a device corresponding to Bob's manager. The <presence> element may contain multiple <tuple> elements, each segmenting a specific device that is contributing to the overall presence of the principal.

Each <tuple> element contains a <basic> element that can have a valid value of either "open" or "closed," corresponding, respectively, to whether the principal is available for communication (i.e., present) or not (i.e., absent).

The last item of interest is the use of namespaces in the XML document. The notification is carrying a payload in the form of a SPIRITS XML document (identified by the Content-Type header). Recall that the SPIRITS schema, as outlined in Appendix A, is extensible through the use of other namespaces; thus, the SPIRITS XML document in Figure 7.5 includes a namespace extension such that elements from a PIDF XML schema can be included in

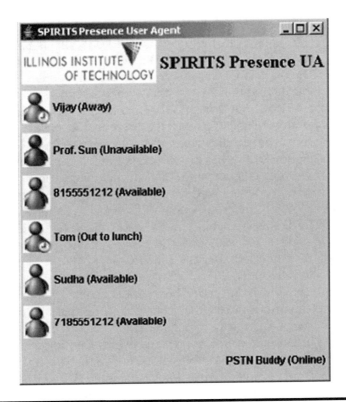

Figure 7.6   Updated presence information.

the SPIRITS XML document. XML's namespace extension mechanism provides a powerful way to represent complex states of an event source.

When Bob's UA receives this notification, it extracts the PIDF information to set the presence state of the device belonging to Bob's manager to "Available" (see Figure 7.6). In this manner, Bob knows in real time when his manager has arrived at the airport. He is thus able to make the best use of his time to prepare for the subsequent meeting. Note that the notification also contained a Cell-ID, which provides rough information on where the manager currently is. We will revisit location-based services in Section 7.3.4.5.

7.3.4.2   Availability

When Bob arrived to work in the morning, he discovered that he has to attend an all-day meeting scheduled at the last minute. He calls his wife to tell her about his plans only to find that his home phone is busy.

Clearly, Bob would like the telephone network to inform him when his home phone becomes available again, so he can call his spouse. The alternative is to keep on trying or have his wife reach him while he is in the meeting. Thus, availability as a service can be applied equally as well to PSTN endpoints. Traditionally, the PSTN knows of the state of devices representing principals, but it has not been able to leverage this information to provide an availability axis to the presence dimension.

To compose the availability status of his home phone line, Bob's UA sends a subscription to the EM that contains the following events drawn from Table 6.1: OAA, OD, TA, and TD. The first two events correspond to the case where Bob's wife was the caller; i.e., she picked up the phone to make a call (OAA) and subsequently disconnected after the call was over (OD). The last two events cover the case if Bob's wife were the target of someone else's call; i.e., she answered the phone (TA) and subsequently disconnected after the call was over (TD).

Because the subscription from Bob's UA reaches the EM while the conversation was already in progress, the EM is unable to create an availability composition; thus, it sets the state of the principal to "Unavailable" as a default (see the last row of Figure 7.7a). For scalability, the EM is stateless, i.e., it does not correlate previous events with future ones. Thus, when the event that rendered to Bob's home line to be busy was published, there was not any pending subscription against it. Thus, it was simply dropped. The statelessness of the EM is explored further in Section 7.4.

Following the acceptance of the subscription from Bob's UA, whenever a OD or TD event is published to the EM, it sends out a notification that

Figure 7.7 Depicting availability.

causes Bob's home line to become "Available," as depicted in Figure 7.7b. Bob can now break out of his meeting to talk to his wife based on the latest availability information imparted to him on an Internet device by the telephone network.

Besides composing simple availability, as shown in Figure 7.7, our system can be used to impart a temporal component to availability as well. For this to occur, the subscription from Bob's UA must reach the EM while there is not already a call in progress on a principal's line. If this is indeed the case, then when the principal picks up the phone to make a call (or receives a call), the EM composes an availability document with temporal information in it. Figure 7.8 shows the result of such an availability composition. Notice the last row: it contains the time since the phone line has been engaged in a conversation.

From a protocol perspective, the information for temporal availability is carried in a PIDF element called <note>. This element, in a PIDF document, contains a string value, which is usually used for a human-readable comment. Figure 7.9 reproduces the XML document that shows that Bob's home line received a call (the published event is TA, which signifies that the terminating party picked up the phone). Also note the

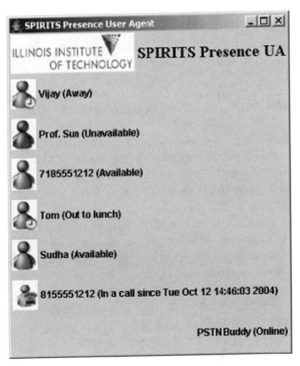

Figure 7.8 Depicting temporal availability.

```
<?xml version="1.0" encoding="UTF-8"?>
<spirits-event xmlns="urn:ietf:params:xml:ns:spirits-1.0"
            xmlns:pidf="urn:ietf:params:xml:ns:pidf">
  <Event type="INDPs" name="TA">
      <CalledPartyNumber>8155551212</CalledPartyNumber>
      <CallingPartyNumber>2015551212</CallingPartyNumber>
  </Event>
  <pidf:presence entity="pres:8155551212@pstn-provider.com">
      <pidf:tuple id="sg891">
        <pidf:status>
          <pidf:basic>open</pidf:basic>
          </pidf:status>
        </pidf:tuple>
      <pidf:note>In a call since Tue Oct 12 14:46:03 2004</pidf:note>
  </pidf:presence>
</spirits-event>
```

Figure 7.9   XML document transporting temporal availability.

PIDF <note> element. When Bob's UA receives such a document, it extracts the value of the <note> element and displays it in the user interface of Figure 7.8.

Figure 7.9 also contains other related information in the form of the calling party's identity (2015551212), which is available to Bob's UA to display to Bob, should it so desire.

7.3.4.3   An IM from the Telephone Network

> Bob is expecting an important call from his brother, which he will likely miss while he is in the meeting. When his brother calls Bob's cellular or office phone, Bob would like the PSTN to inform him through an unobtrusive IM on his PDA.

To do so, Bob runs a UA on his PDA that he will be taking with him to the meeting. His UA asks him for phone numbers to monitor (see Figure 7.10). Bob enters two phone numbers, corresponding to the numbers of his cellular phone and his desk phone. Bob is interested in getting a notification as soon as someone calls him at either of the numbers. Bob's UA sends out a subscription with the filter containing events for the two phones Bob uses; Figure 7.11 shows such a subscription. There is one salient point in Figure 7.11. Note the Accept header. This header contains three SIP requests that Bob's UA can accept: SUBSCRIBE, NOTIFY, and MESSAGE. Of interest to us is the last request, MESSAGE. This is a SIP extension, defined in [CAM02a], which transmits discrete text messages

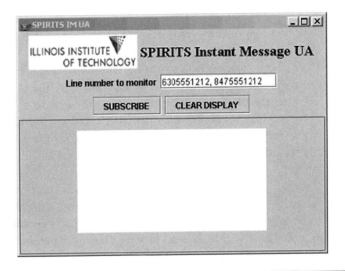

Figure 7.10   User interface for the IM user agent.

```
SUBSCRIBE sips:bob@pstn-provider.com SIP/2.0
...
Allow: SUBSCRIBE, NOTIFY, MESSAGE
Allow-Events: spirits-user-prof, spirits-INDPs
Event: spirits-INDPs
Accept: application/spirits-event+xml, text/plain
Content-Type: application/spirits-event+xml
Content-Length:...

<?xml version="1.0" encoding="UTF-8"?>
<spirits-event xmlns="urn:ietf:params:xml:ns:spirits-1.0">
    <Event type="INDPs" name="TAA" Mode="N">
      <CalledPartyNumber>6305551212</CalledPartyNumber>
    </Event>
    <Event type="INDPs" name="TAA" Mode="N">
      <CalledPartyNumber>8475551212</CalledPartyNumber>
    </Event>
</spirits-event>
```

Figure 7.11   Subscription for an instant message.

using SIP as the transport protocol. The body of such text messages is defined by the MIME-type text/plain, which is included as an accepted payload in the Accept header of the request.

The payload of the SIP SUBSCRIBE request shown in Figure 7.11 is a SPIRITS XML document. The filter represented by the XML document, in essence, informs the PSTN that a notifier is willing to receive notifications when the TAA event is published for two phone numbers (6305551212 and 8475551212, corresponding to Bob's cellular phone and Bob's desk phone, respectively).

At some later point in time, Bob's brother calls Bob's desk phone. This causes the selection process in the EM to execute the filter installed in Figure 7.11. The result of this will be the publication and dissemination of the TAA event to Bob's UA. When Bob's UA receives the notification, it displays an IM in the user interface, as depicted in Figure 7.12. The IM discreetly informs Bob that his brother attempted to reach him by calling Bob's desk phone at a certain time.

There is a subtle protocol interplay occurring behind the scenes to fulfill this service. Note that when Bob's UA issued a subscription, it indicated support for transporting IM messages in SIP through its acceptance of the MESSAGE request. The EM thus has a choice in informing Bob's UA of the event. It can choose one of three choices: First, it can send a NOTIFY message, followed by a MESSAGE request. The former informs Bob's UA of the event that occurred, and the latter contains the IM. Second, it can send one notification message that contains a multipart MIME payload. Multipart MIME is an IETF standard [FRE96b] that allows multiple objects, each corresponding to a specific MIME type, to coexist in a single payload. Thus, the single payload will contain two parts: (1) the SPIRITS XML document carrying the event that occurred and (2) a plain text message containing the IM. The third choice the EM has is to simply send a NOTIFY message containing a

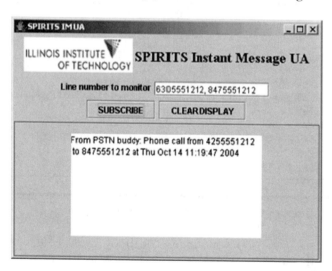

Figure 7.12 Incoming call notification as an instant message.

SPIRITS XML document and depend on Bob's UA to construct an IM from the contents of the SPIRITS document and display it to Bob.

In our implementation, we chose the second option, i.e., a NOTIFY followed by a MESSAGE. Figure 7.13 contains portions of a MESSAGE request, and Figure 7.14 contains a flow of the messages between the communicating entities.

Finally, a feature of our implementation is the support of a pseudo-principal called the PSTN Buddy. Figure 7.6 to Figure 7.8 depict this pseudo-principal, which is always present (see lower right-hand side of the user interface in these figures). This pseudo-principal is used by the telephone network to originate instant messages; such messages arriving at a UA are shown to be from the PSTN Buddy (see the IM in Figure 7.12 for an example).

```
MESSAGE sips:bob@isp-provider.com SIP/2.0
...
Content-Type: text/plain
Content-Length:...

From PSTN buddy: Phone call from 4445551212 to
8475551212 at Thu Oct 14 11:19:47 2004
```

Figure 7.13   MESSAGE request.

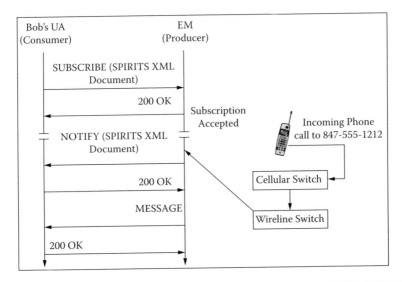

Figure 7.14   Message flow.

7.3.4.4 Transforming an SMS to an IM

Bob keeps in touch with his colleagues at other locations through the cellular SMS. This morning, Bob did not bring his cell phone to work, so he will miss all SMS messages destined to his cell phone. Bob would like the PSTN to intercept the SMS messages destined for his phone and transform them to IMs to be delivered to another of Bob's devices connected to the Internet.

This is a good example of an application-specific service. To realize such a service, an event filter must be created that contains an event pertinent to the operation of an SMS, especially an event that is published when the cellular network attempts to determine Bob's location so it can send him the SMS. Of all the events we have cataloged thus far, none of them is amenable to use in the SMS application. Yet, clearly, the cellular network knows if the recipient of an SMS is unavailable because it queues the SMS for later delivery. The challenge is getting access to this trigger and representing the event filter in a schema pertinent to SMS.

We propose an SMS XML schema, provided in Appendix C, to solve the problem of representing SMS-specific events. Regarding access to a trigger that is appropriate for publishing the SMS arrival event, we have two choices. First, we can detect this trigger in a specialized server called a message center (MC) [GAL97]. The MC provides a store-and-forward function for SMS messages. An MC can serve as a good event source of SMS-related messages.

Another equally good event source could be the mobile switching center (MSC) itself. The MSC is involved with all aspects of signaling in the cellular network. For instance, it knows whether the recipient of the SMS message is registered with the network. If the recipient is not registered, the MSC can take appropriate action by acting as an event source for SMS-related triggers.

In our implementation, we used the latter approach. When the MSC determined that the recipient was not responding to a paging message, the MSC, acting as an event source, sent an SMS-related event to the EM. The contents of the message included the sender's tel URI and the SMS message itself. The SMS message was transmitted to the MSC using the SMS simulator in the laboratory.

Figure 7.15 contains an XML filter that Bob's UA may use to subscribe to the arrival of an SMS message. This document identifies the principal (Bob) through his cellular phone number (6305551212). The filter establishes two constraints identified by the <DeliveryType> element. The first constraint (Failure) signifies failure to locate Bob; i.e., the cellular network could not locate Bob, possibly because Bob's cellular phone is turned off. In such a case, the SMS should be transformed to an IM message and sent to the URI provided in the <IM> element.

The second constraint (In-addition-to) specifies that even if Bob is successfully located, the cellular network is to send the SMS to his device and, in addition to that, transform the SMS into an IM and deliver it to the URI provided in the <IM> element.

The URI in the <IM> element has a parameter (method=MESSAGE), which in SIP signifies the name of the SIP request that should be used to contact the URI. That is, the cellular network will use the SIP MESSAGE extension to transport the IM.

Figure 7.16 contains a screen shot of our integrated user interface that depicts three smart-space services: the left-most panel contains a presence and availability service (already discussed in Sections 7.3.4.1 and 7.3.4.2,

```
<?xml version="1.0" encoding="UTF-8"?>
<sms xmlns="http://www.iit.edu/sms-1.0"
    xmlns:xsi="http://www.w3.org/2001/XMLSchema-instance"
    xsi:schemaLocation="http://www.iit.edu/sms-1.0 SMS.xsd"
    Principal="tel:6305551212">
      <DeliveryType>Failure</DeliveryType>
      <DeliveryType>In-addition-to</DeliveryType>
      <IM>sips:bob@iit.edu;method="MESSAGE"</IM>
</sms>
```

Figure 7.15   An XML document with a filter for converting SMS to an IM.

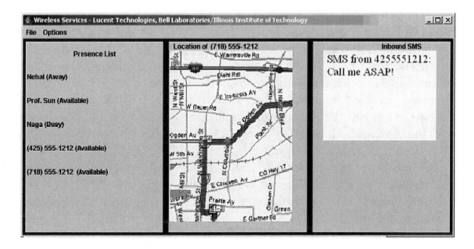

Figure 7.16   An integrated user interface.

respectively), the center panel contains geolocation-based services (discussed in the next section), and the right-most panel contains a text area that displays the incoming IM equivalent of an inbound SMS message destined to the recipient's cellular phone. In the SMS panel of Figure 7.16, Bob's brother, identified by his device (4255551212), has sent an SMS message to Bob's cellular phone. When the MSC receives the SMS message, it acts as an event source and sends the message to the EM. The EM extracts the SMS text from the message, creates a SIP MESSAGE request with the Request-Uniform Resource Identifier (R-URI) set to sips:bob@ip-provider.com, and transmits the message securely to Bob's UA.

Our architecture and methodology also enable multimedia messages (MMSs) [3RD02] to be delivered to a principal using only a 2.5G phone. The principal would install a filter with the MMS center such that MMSs are shunted to his Internet UA, while normal SMS messages go to his 2.5G phone (or his Internet UA). MMS is a technology that extends SMS into the space of multimedia. Instead of simply sending text messages, MMS allows its users to send messages in the form of images, audio, video, text, and combinations of these.

7.3.4.5 Location-Based Services

> Bob would like to track the location of his manager as she makes her way to Bob's office from the airport. This will allow Bob to better prepare for the meeting by taking advantage of any extra time afforded to him if the manager gets delayed by traffic.

When Bob's manager arrived at the airport, Bob knew of this because he had subscribed to the presence state of his manager through the manager's device. Because Bob would like additional, real-time information on the progress of his manager as she drives over, Bob instructs his UA to install a new subscription filter with the cellular network. This subscription filter, depicted in Figure 7.17, requests notification for two events associated with the manager's device: LU7SV and LUDV (i.e., location updates in the same area and a different area).

As Bob's manager travels, she crosses from one cell area to the next. The cellular system thus keeps track of her progress. Whenever her location changes in this manner, the MSC receives a location update message. Technically speaking, the detection of a principal in a new serving system is an example of a registration event [GAL97, pp. 162–163].

Registration always occurs when the cellular phone is turned on. Timer-based or autonomous registration occurs at periodic intervals — ranging

```
SUBSCRIBE sips:7185551212@provider.com SIP/2.0
...
Event: spirits-user-prof
Content-Type: application/spirits-event+xml
Content-Length:...

<?xml version="1.0" encoding="UTF-8"?>
<spirits-event xmlns="urn:ietf:params:xml:ns:spirits-1.0">
    <Event type="userprof" name="LUSV" mode="N">
      <CalledPartyNumber>7185551212</CalledPartyNumber>
    </Event>
    <Event type="userprof" name="LUDV" mode="N">
      <CalledPartyNumber>7185551212</CalledPartyNumber>
    </Event>
</spirits-event>
```

Figure 7.17   An XML document with a location filter.

from ten minutes to one hour — while the cellular phone is turned on. The granularity of autonomous registrations is typically transmitted to the cellular phone by the serving MSC. And finally, if a principal of the cellular phone is engaged in a conversation, the system tracks the location of the principal for handoffs (a handoff is the seamless transfer of an ongoing call from one base station to another). The end result of the registration and handoff process is that the MSC is aware of the location implications of the cellular phone.

In our laboratory setup, we employed the signal attenuator to depress the signal strength in one cell area and increase the intensity in an adjoining cell area. The cellular phone would then register with the system in the new cell area, thus simulating mobility. When the MSC received a message from the cellular phone that involved such movements, it, in turn, published an event to the EM containing this information.

The EM executed the selection process to send out the event to an Internet UA. In the notification, the EM included a Cell-ID parameter (as required by Table 6.3). We programmed the UA to use the Cell-ID as an index into an associative array of images; each image corresponded to the geographical area covered by that cell. The Internet UA extracted the Cell-ID parameter from the notification and used it to retrieve a map image that was rendered on the user interface (see center panel of Figure 7.16).

Tracking the location of a principal in this manner is useful, but probably not very optimal. For one, cellular boundaries may change, requiring updated image maps to be downloaded. Furthermore, the granularity of the autonomous registrations may be fairly large, which makes such tracking inadequate. And finally, the geographic area encompassed by a cell may vary; in dense urban canyons, a cell may be fairly limited in diameter, but in sparsely populated rural areas, a cell may be defined by a larger boundary. However, despite these shortcomings, location tracking in the manner we describe is useful for a variety of operations. Consider, for example, fleet tracking. A taxi dispatcher needs to know the approximate location of a taxi nearest to the customer when a call arrives at the dispatcher. Or, as a further example, parents may consider using this technology to know the approximate whereabouts of their children. Some other examples include targeted advertising, wherein a business solicits customers that are in the vicinity through an SMS message (or an IM, if the customer's phone is connected to the Internet), and a friend finder, where two friends in the same geographic area (a mall, for instance) are notified of each other's proximity. In all of these examples, an approximate location instead of an exact location suffices, and the cellular network already has this information.

Exact locations can be provided by the geographical positioning system (GPS). However, this technology has its drawbacks, primary among them that it does not always work indoors. A cellular signal, on the other hand, does not suffer from this drawback.

7.4 Design and Implementation of the Event Manager

In event-driven, producer–consumer-oriented middleware, two dimensions are usually considered fundamental [CUG01]: the expressiveness of the subscription language and the architecture of the event dispatcher. Both of these dimensions are pertinent to the EM, our event dispatcher.

The expressiveness of a subscription language is characterized by *subject*-based systems, where subscriptions identify only classes of events belonging to a given subject, and *content*-based systems, where subscriptions contain expressions to allow sophisticated matching on the event content [CUG02]. The former systems are simpler, whereas the latter are bounded by the complexity of the expression when evaluating events for millions of users with different interests. Our subscription language is fairly simple and can be characterized as subject based.

The architecture of an event dispatcher can be either *centralized* or *distributed* [CUG01]. In the former case, a single component acts as an event dispatcher, and in the latter case, a set of interconnected dispatching servers cooperate in collecting subscriptions arriving from consumers and in routing events. The disadvantages of the distributed architecture include figuring out strategies to route subscriptions and events and to maintain a minimal spanning tree that includes all dispatching servers. The advantages, of course, are reduced network load and increased scalability.

Our use of the EM in our architecture is marked mostly by the centralized model. Due to the security concerns outlined in Section 7.3.2, we assume that the EM that received the subscription from the consumer will also send the notification to it. However, as is the case with most telecommunication software and hardware, redundancy is implemented by maintaining an inactive mated pair, and scalability is addressed by running multiple EMs. To the consumer, they appear as one EM, but for scalability reasons, there will be more than one.

7.4.1 Design of the EM

The EM is a critical piece in our architecture. All events published in the system by multiple event sources arrive at the EM, where they are mediated. Mediation consists of running a selection process on the event to determine if a consumer is interested in receiving a notification of that class of event. If so, the EM constructs a SIP notification request and dispatches it toward the consumer in a secure fashion. If the selection process did not result in an interested consumer, the event is simply discarded.

For the advantages it affords us, we designed the EM to be stateless. An arriving event is not influenced by past events, and it will not have any influence on the arrival rate of future events. The stateless property of the EM affords us three incentives. First, for performance analysis, we can simply treat it as a Poisson process with exponentially distributed interarrival time. Second, the stateless property helps in fault tolerance, because a backup EM can effortlessly step in if the primary EM fails, as well as scalability, because an incoming event can be distributed to one among a set of EM servers. Finally, being stateless implies that multiple EMs can be started to share the load; any EM can service an incoming event distributed to it.

Figure 7.18 contains a logical design of the EM. The bottom layer is composed of an RFC3261 [ROS02]-compliant SIP transaction manager that we developed [ARL04]. Using the services of the SIP transaction managers are two engines: a subscription engine and a notification engine (more details on these engines are provided below). The SIP transaction manager supports SIP over TCP, User Datagram Protocol (UDP), and TLS. It parses incoming SIP requests and hands them to the subscription engine, and accepts outgoing messages from the notification engine and uses SIP routing techniques [GUR01] to deliver the notifications to consumers.

Sitting above the transaction manager and using the services of it are two engines: a notification engine and a subscription engine. The subscription engine accepts subscription requests from the consumers containing event filters, authenticates the consumers, and processes the event

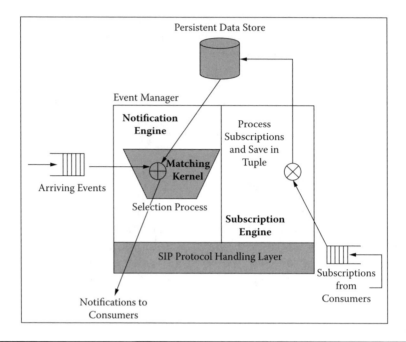

Figure 7.18   Design of the Event Manager.

filters by updating the policy tuple stored in the database. When the policy tuple was initially created, the C-URI_α element was left null, as was the C-Ev_α list (see Section 7.3.3). After a subscription has been accepted, both elements are populated as follows: The phone number of the principal is extracted from the XML document and used to find a matching tuple from the database. Once found, the C-Ev_α list is populated from the discrete events present in the XML document of the subscribe request. The C-URI_α element is populated either from a special field of the SIP subscribe request called a Contact header or, depending on the schemas supported by the EM, from the XML document itself (recall that the SMS schema outlined in Appendix C inserts the URI where the notification should be sent in the XML document). The end result of an accepted subscription is that the retrieved tuple is now complete, i.e., all the elements have been assigned appropriate values.

The subscription engine parsed the XML payload using expat v1.95.5, an XML parser that can be downloaded under an open source from http://sourceforge.net/projects/expat. Expat is used extensively in other open-source software that demand a fast parser, including the software that implements the Python and Perl languages. Expat is a nonvalidating parser, which means that it validates an XML payload to ensure only that it is well formed. It parses and returns elements, attributes, and text from the XML payload. It does not validate the payload; i.e., it does not ensure that the values of attributes are legal, or that the elements belong to a certain namespace. There exist validating XML parsers that do all of this, but as can be expected, validation comes at a cost of parsing speed. Dynamic validation of XML documents implies more processing time per document, and thus a smaller number of documents processed per unit time. Therefore, we chose a nonvalidating parser, but before running the experiments, all our XML payloads were validated by a software-based XML editor to ensure that they were well formed and that they were valid according to their respective schemas.

The notification engine contains a queue where published events are deposited by the multiple event sources. Events are serviced on a first-come first-served basis. The notification engine contains the selection process, exemplified by a matching kernel. The matching kernel is the core of the notification engine. It implements the matching algorithm (referred to in previous parts of this chapter as the *selection process*) for the subscription language and publication model deployed. The matching kernel is implemented as a C interface to the SQLite database. The clear advantage of this approach is in the features of a database manager that we acquire with minimal effort. We do not have to redevelop solutions for the well-known problems of inserting subscriptions so that they can be retrieved quickly, indexing the subscriptions, transaction integrity, and the like.

Our subscription language, owing to the decision to use a database manager, is thus very simple. An incoming event contains a set of attributes, a subset of which are matched (exact match) to the subset of attributes of the policy tuple, P_α, stored in the database (see Section 7.3.3). An incoming event therefore acts as a constraint on the tuples in the database to derive a match.

An incoming event is represented as a constraint $\phi = \{ev_\phi, P_\phi, evp_\phi^1, ..., evp_\phi^n\}$, where ev_ϕ is the published event, P_ϕ is the event source, and $evp\phi^1, ..., evp_\phi^n$ are parameters related to ev_ϕ. The matching kernel will return success if and only if

$$P_\phi = Pr_\alpha \wedge ev_\phi \in Ev_\alpha$$

Once a match has been determined, the notification engine sends a notification to the consumer identified by the URI in the $C\text{-}URI_\alpha$ element. The notification consists of a payload that includes a SPIRITS XML document. Attributes $evp_\phi^1, ..., evp_\phi^n$ pertinent to the event ev_ϕ are included in this document. Depending on the XML schemas supported by the EM, the notification engine may also perform additional services, such as sending out an IM. However, in the interest of processing speed, the set of such services must be limited. If the number of such ancillary services increases, it may become necessary to introduce a service engine, which will offload this task from the EM.

Although Figure 7.18 contains the logical design of the EM, the EM we implemented was realized as two threads running in a process. One thread implemented the functionality required of the subscription engine, and the other thread implemented the role of the notification engine. The notification engine contained the matching kernel. Both threads accessed the database to store, retrieve, and update the policy tuples. Data integrity and concurrency were provided by the underlying database manager.

7.5 Performance Analysis of the Event Manager

To derive a performance model of the EM, we focused on the notification engine. Certainly, the subscription engine could have been used, but the processing performed by it is less intensive than the processing that the notification engine undergoes on the arrival of an event. The subscription engine retrieves subscriptions from consumers from its queue, parses the SIP request and the payload, updates the database, and sends a response to the consumer. The notification engine, on the other hand, must retrieve published events from its queue and execute the selection process to determine if a consumer is interested in receiving a notification. If so, it transmits a notification toward the consumer and awaits a response indicating that the notification arrived at its destination.

7.5.1 Assumptions and Realities

For performance analysis, we turned off the authentication module in the notification engine. Encryption and authentication introduce latency in the system. Depending on various factors when encryption is used, including the choices of an asymmetric algorithm, a private key's size, a symmetric algorithm, a digest algorithm, whether session resumption should be used, and the size of records exchanged, throughput suffers considerably [RES01]. Simulation studies have demonstrated that as the number of active clients using encryption increases, the queue size of the HTTP server servicing the clients increases dramatically [RES01, pp. 202–204].

We created and inserted 1 million policy tuples in the SQLite database. Each tuple represented a policy installed by a principal of the system. The subscription engine updated the policy tuple based on a subscription request from a consumer. The notification engine used an incoming event as a constraint in the selection process to determine if a consumer was interested in the event.

We executed our experiments on a Sun Microsystems Netra 1405 UltraSPARC-II, four 440-MHz CPUs, with 4 GBytes of main memory running the Solaris 5.9 operating system. As mentioned previously, our focus was on the performance of the notification engine. We simulated event arrival at the notification engine following a Poisson distribution with a mean arrival rate (λ) that we increased across three successive runs: 200, 400, and 600 events/s. The notification engine retrieved the event from its queue, parsed it, executed the selection process on it, and sent the notification to the consumer using the URI stored in the policy tuple. All incoming events resulted in the selection process returning a match; i.e., all events led to a notification. We ran a consumer on the same host that was running the notification engine, but on a different port.

7.5.2 Determining Service Time per Event

Table E.1 contains the raw data of our executions and shows the total execution time and the average service time per event. The total execution time is the time it took for the last of the SIP 200 OK responses from the consumers to arrive at the notification engine. Because the consumers were on the same host as the notification engine, the loopback interface was used to send notification requests over the UDP transport. Responses arrived over the same loopback interface using UDP. Of interest to our analysis is the average service time per request.

We characterize the average service time at the notification engine as $E[S]$:

$$E[S] = t_p + t_d + t_n \tag{7.1}$$

where:
t_p = The time required to retrieve a pending event from the queue and parse it
t_d = The time required to execute the selection process
t_n = The time required to send the notification to the consumer

Note that we have not included the time to receive a 200 OK for the notification in the average service time (we do so for the total execution time of the process). To a great extent, this time will be bounded by the network traffic and traffic dynamics between the notification engine and the consumer — factors that we cannot control. In our experiments, we normalized this delay by using the loopback interface. However, to err on the side of caution, the average service time does not include message propagation delay, even over the loopback interface. In Equation 7.1, t_n only includes the time required by the notification engine to place the notification request on the queue of the SIP transaction manager. It

Table 7.2   1/μ and ρ per Event at Different Arrival Rates (λ)

	$\lambda = 200$	$\lambda = 400$	$\lambda = 600$
1/μ (ms)	1.42	1.35	1.34
ρ	0.28	0.54	0.80

Note: $S = 1/\mu$, average service time per event (ms); λ, arrival rate (events/s); ρ, traffic intensity ($\rho = \lambda/\mu$).

specifically does not include the time the SIP transaction manager spends in processing and transmitting the notification request.

Table 7.2 summarizes the average service time per event from Table C.1. It also includes the measure of traffic intensity (ρ).

Figure 7.19 shows the results graphically: (a) plots the total execution time at increasing arrival rates, (b) plots the service time across the different arrival rates, and (c) plots the traffic intensity across different arrival rates. Figure 7.19a indicates a constant increase in total execution time, which is to be expected. As the offered load, λ, to the system increases, it takes a longer time for each such notification to be processed, transmitted to the consumer, and result in the receipt of a 200 OK message from the consumer (recall that the total execution time includes the receipt of a 200 OK from the consumer).

Figure 7.19b plots the service time per event as the offered load to the system increases. We note that the service time remains somewhat constant, despite the increase in the load. This suggests to us that the machine running the experiments is powerful enough to keep up with the offered load.

Figure 7.19c plots the traffic intensity, ρ, as the offered load to the system increases. As can be observed from Table 7.2 and Figure 7.19c, ρ increases with a corresponding increase in λ. In traditional queuing analysis theory [JAI91, KLE75], stability concerns dictate that $\rho < 1.0$; thus, we halted the experiment at a mean arrival rate of 600 events/s, or 2.16 million events/h.

7.5.3   Calculating Blocking Probability: Erlang-B Analysis

In a system such as the notification engine, the effects of queuing have to be minimized. In other words, to maintain the real-time nature of notifications, a published event must spend as little time as possible in the queue waiting to be serviced; if it spends too much time in the queue, the impact of the information it conveys is lost. Thus, the number of event managers should be such that the blocking probability is very low, say 0.01 or 1 percent of all events coming in find all the resources busy.

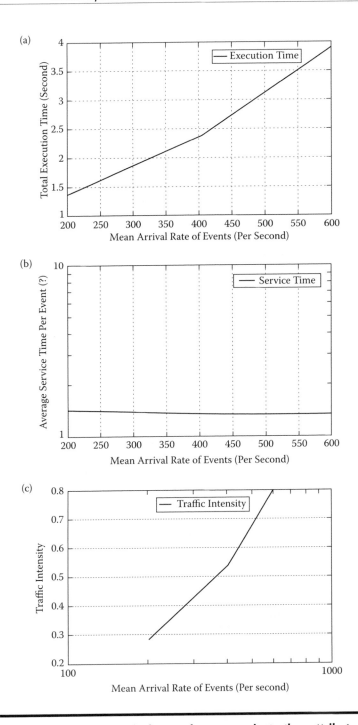

Figure 7.19 Plots of mean arrival rate of events against other attributes. (a) λ versus total execution time. (b) λ versus 1/μ. (c) λ versus ρ.

Two types of analysis yield blocking probabilities: Erlang-B and Erlang-C. Erlang-B is used when failure to get a free resource results in denial of service; i.e., if an event source publishes an event, and there are not any resources to process that event at the notification engine, then the event will be dropped. Erlang-C is useful if what is modeled is the probability that the event will be queued until it is processed by the notification engine.

Between these two analyses, we analyze our system using Erlang-B. Because the events coming into the system have a temporal quality associated with them, i.e., the consumer needs to be notified immediately, the system is ill-served if the events experience excessive queuing delays. Erlang-B analysis allows us to analyze the system such that it is dimensioned for a low blocking probability, i.e., very few events are to be dropped due to resource exhaustion.

Given an infinite source of event generation and a Poisson arrival rate, the Erlang-B formula can be expressed as [KLE75]

$$B(c, \rho) = \frac{\dfrac{\rho^c}{c!}}{\displaystyle\sum_{i=0}^{c} \dfrac{\rho^i}{i!}} , \text{ where } \rho = \lambda/\mu \qquad (7.2)$$

$B(c, \rho)$ describes the fraction of time all c servers are busy at a traffic intensity of ρ. Table 7.3 shows the number of event managers we would need if we wanted to keep the blocking probability to 0.000001 (i.e., one event of a million experiences resource contention and is dropped).

7.5.4 Modeling the Event Manager as an M/D/1 Queue

For a traditional queuing analysis of the system, we model the EM as an M/D/1 queue.

Table 7.3 Number of Servers (c) Needed for Various B(c, ρ)

$B(c, \rho)$	$\rho = 0.28$	$\rho = 0.54$	$\rho = 0.80$
0. 000001	$c = 6$	$c = 8$	$c = 9$
0. 00001	$c = 6$	$c = 7$	$c = 8$
0. 0001	$c = 5$	$c = 6$	$c = 7$
0. 001	$c = 4$	$c = 5$	$c = 6$

Recall that the EM is stateless to begin with; thus, we model it as a Poisson process with exponentially distributed interarrival time. Because we have experimentally measured the time required to service one event, we will use a deterministic service time distribution. Using standard approaches to M/D/1 queues [JAI91], the parameters of interest are as follows:

λ: Arrival rate in events per second

$E[S] = k$: Constant service time, from Equation 7.1

$\mu = 1/E[S]$: Service rate

$\rho = \lambda E[S]$: Traffic intensity

$E[N] = \rho + \rho^2/2(1 - \rho)$: Mean number of events in the system, or queue length

$E[n_q] = \rho^2/2(1 - \rho)$: Mean number of events in the queue

$E[r] = E[S] + \rho E[S]/2(1 - \rho)$: Mean response time; includes time waiting for service and time receiving the service

$E[w] = \rho E[S]/2(1 - \rho)$: Mean waiting time, i.e., the time difference between the arrival time and the instance the event starts to receive the service

Figure 7.20 contains two graphs that plot the values of our analysis as the arrival rate varies from 200 to 600 events/s. Figure 7.20a shows the increase in the mean number of events in the system ($E[N]$) and the mean number of events in the queue ($E[n_q]$) as the arrival rate is increased. Figure 7.20b plots the mean response time ($E[r]$) and mean wait time ($E[w]$) as λ increases. From Figure 7.20a, even when the arrival rate is the highest at 600 events/s, the mean number of events in the system is 2.5 events and the mean number of events waiting in the queue to get services is only 1.6 events. Likewise, at $\lambda = 600$ events/s, the mean response time is 0.004 ms and the mean wait time is 0.002 ms.

Results of M/D/1 analysis in the context of s servers (M/D/s) have been tabulated for numerous cases in [HIL81]. Figure 7.21 shows the tabulation of $E[n_q]$ across s servers (in the figure, $E[n_q]$ is labeled "L" on the y-axis). For s servers, ρ is

$$\rho = \lambda/s\mu \qquad (7.3)$$

From Table 7.3, we know that at a traffic intensity $\rho = 0.80$, we will need seven servers to keep the blocking probability B(c, ρ) at 0.0001 (i.e., one of 10,000 events is dropped).

At an arrival rate $\lambda = 600$ events/s and a service rate $\mu = 746$ events/s, Equation 7.3 yields $\rho = 0.11$ for $s = 7$ servers. The corresponding lookup of $\rho = 0.11$ in Figure 7.21 yields $E[n_q] = 0.6$ events queued in the system

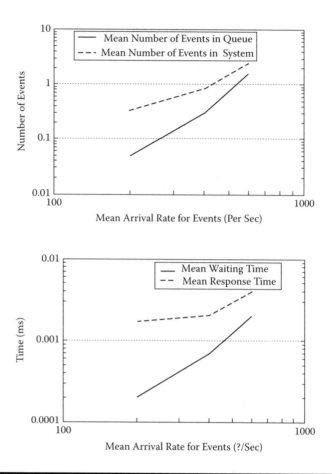

Figure 17.20   Plots from M/D/1 analysis. (a) λ versus *E[N]*, *E[n_q]*. (b) λ versus *E[r]*, *E[w]*.

if seven servers are processing events. Our earlier M/D/1 analysis shows that for the traffic intensity ρ = 0.80, $E[n_q]$ = 1.6 events (i.e., 1.6 events are queued in the system) if one server processes all the events. With seven servers, utilization per server is fairly low; hence, the queue does not build up too drastically.

7.6   Related Work

Although pervasive computing itself is an area of active research, its application to the telecommunications domain to provide a wider communication experience is a fairly recent phenomenon. Efforts exist in mobility management for pervasive computing [KAN01, KAN02, PAN02],

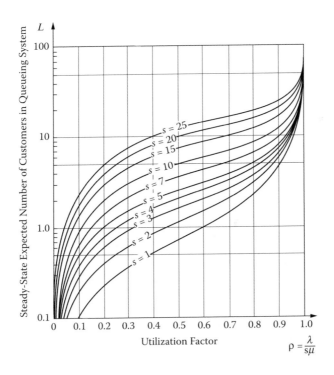

Figure 7.21   Values for E[n_q] for the M/D/s model. (Reproduced from Figure 16.11 in Hillier, F. and Lieberman, G., *Introduction to Operations Research*, 5th ed., McGraw-Hill, New York, 1990.)

but unlike our work, these do not take into account the PSTN events as a stimulus for providing pervasive communications, and they do not enable the adaptation aspect of our architecture (discussed in Section 7.7). Furthermore, existing literature views pervasive computing in the telecommunication domain within the context of an end-to-end IP network [KAN01], more specifically, a 3G all-wireless IP network. We feel that although this is indeed a laudable goal, the current state of telecommunication is such that the wireless circuit-based infrastructure (2G, 2.5G) will not disappear any time soon. To that extent, our architecture successfully exploits the current circuit-based infrastructure and, indeed, merges it with the Internet to provide a pervasive computing platform to its users.

Additional work has been done in context-aware communication, where the changing information about an individual's location, environment, and social situation can be used to initiate and facilitate people's interactions with one another or in a group [COL01, SCH02, TAN01, WAG03]. However, the main thrust of these works is on group communication. The availability of each member in the group is aggregated by interfacing with the myriad devices used by the members (computers, PDAs, etc.), as well as querying online calendar and appointment managers. These systems provide a convenient clearinghouse type of functionality by aggregating all the means by which an individual can be contacted (phone, e-mail, instant message, etc.). Their usage of the PSTN is limited to placing phone calls through it. They do not take into account the powerful call models and wireless infrastructure of the PSTN to generate events in the same manner that our work has demonstrated.

To the extent that our architecture enables presence-based services, it can be contrasted with similar systems that exhibit awareness, such as Sun Microsystem's Awarenex [TAN01] and Milewski et al.'s Live Address Book [MIL00]. Awarenex does indeed designate if a cellular phone is in the middle of a conversation, but it does so in an incremental and *ad hoc* fashion that mandates that the call request be placed through an Awarenex server. Only if this is done does the system provide real-time status updates. It is easy to bypass the server completely. Our architecture arranges for presence-related events to be detected by the deployed and pervasive cellular network. Furthermore, to allow richer Internet services, our architecture provides for many more events beyond those required for real-time status updates.

The Live Address Book also permits its users to provide real-time updates of the status of their phones, but it does so manually. As Milewski et al.'s research indicates, users will not always consciously remember to update their status. Our architecture, by contrast, updates the status automatically.

Stanford's Mobile People Architecture (MPA) is another effort to bridge the wireless and Internet networks [MAN99]. However, its main goal is to route communications to a mobile person, independent of the person's location communication device being used. MPA's goal differs from our work, which aims to provide discrete events to user agents on the Internet for service execution.

The Parlay Group (www.parlay.org) is an industry consortium that specifies application programming interfaces (APIs) to integrate telecommunications network capabilities with arbitrary applications. It is paramount to note that Parlay specifies a programming interface only, not a communication protocol. The work described in this paper could serve as an "on the wire" protocol underneath the Parlay APIs.

Commercial enterprises like Yahoo! allow a cellular phone to become a buddy in a presence list. However, this feature is only provided for phones that are connected to the Internet and is not integrated with call processing. As we have demonstrated, our architecture mitigates both of these shortcomings. A principal using a cellular phone can participate in the buddy list of any presence server (assuming, of course, that the presence protocol implemented by the vendor of the presence server is open). Furthermore, our architecture also eliminates service isolation by integrating call-related knowledge in disseminating availability information as well.

7.7 Conclusion

In this chapter, we have presented a set of discrete services that collectively create a smart space in the telecommunications domain. We presented an implementation of the smart space that demonstrated the feasibility of the application of this work to pervasive computing. We have done this in a manner that is conducive to the use of other protocols (SIP) as well as compatible with the semantic and ontological representation of other information on the Internet (a SPIRITS XML document can contain PIDF elements, for instance).

In Section 7.2, we outlined the four research thrusts in pervasive computing [SAT01]. Clearly, our work demonstrates the first thrust: effective use of smart spaces. We consider the telephone network and the Internet to be two disjoint worlds. Our architecture allows these disjoint worlds to synergistically come together and enables the sensing and control of one world by another. The second thrust is invisibility. Although we do not claim complete disappearance of pervasive computing technology from the user's consciousness, the architecture presented does achieve minimal user distraction. The users will have to configure their user agents so that they issue subscriptions for interested events, but after that the agents become nonintrusive. For some services (like enabling presence on a PSTN phone), the owner of that phone line may have to acquiesce to being monitored in this fashion and, indeed, may have a list of preferred users who can so monitor her. But beyond that, no further intrusion is necessary.

Localized scalability is the third thrust. Good system design can aid in increased scalability by limiting the interaction between the distant entities. This is especially relevant for wireless networks with their pronounced bandwidth and energy constraints. Our system achieves localized scalability by parsimonious use of message exchange between the consumers and producers: at the minimum, two messages are exchanged. One message is exchanged from the consumer to the producer subscribing to events of interest, and a second one from the producer to the consumer notifying it of the action. Furthermore, localized scalability is also enhanced by the event-based architecture; instead of a consumer constantly polling the producer, the consumer is simply notified when the event of interest is published. And finally, localized scalability is also enhanced by the algorithm we presented in Figure 6.8. If bandwidth is at a premium, this algorithm ensures that no more than one event of the same type is published to the consumer in a 15-second time interval.

The final thrust is masking uneven conditioning. We believe that this thrust — possibly accomplished by having personal computing space provide a canonical representation of an environment — is closely coupled with *adaptation* [SAT04], or automatically adjusting behavior to fit circumstances. Our architecture allows for the pervasive communication services to adapt themselves in such a way that an atomic service like presence can be applied to all types of entities: physical users and inanimate phone lines. Similarly, there is no reason that an automaton (the telephone network) cannot send an instant message to a physical user; our architecture demonstrates the needs and scenarios for this to happen. Yet another example of adaptation is how an SMS message turns into an IM to be delivered on the Internet.

Thus far, the two networks have been isolated from each other as far as service sharing is concerned. Certainly, the PSTN has been able to tell whether a device was busy or not, but it has not been able to impart this information to others. The good that imparting this information will do has been documented by many researchers [COP01, LUC04, MON01]. Our architecture and implementation demonstrate the feasibility of synergistically coupling the networks at the services layer. Doing so also solves the service isolation problem we outlined in Chapter 6.

Chapter 8

Conclusions

Over the course of their respective lifespan, the public switched telephone network (PSTN) and the Internet have evolved in divergent ways. The PSTN has been characterized by centralized control of network resources where the intelligence required to process the information resides in the core of the network. The Internet espoused a view that was completely opposite. The network itself was simply considered a transport to move information in the form of bits toward the edges, where more powerful and intelligent machines operated on these bits. Not surprisingly, their respective service architectures reflect this division in philosophy.

PSTN services, owing to their simple endpoints, reside in the core of the network and are executed by trusted entities. Internet services, on the other hand, have the potential of being executed by network intermediaries, but are by and large controlled by the far more powerful endpoint, which may not trust the intermediaries. Another artifact, given the manner in which the service architectures for these networks have evolved, is the degree of personalization. PSTN services are characterized by personalization of the service to the principal, i.e., the call waiting service is applicable to the principal receiving the call. The principal making the call has no way of indicating a preference for this service and, indeed, may not even know that the principal being called is busy. Certainly, Internet services allow personalization as well, but they are marked by an additional dimension: information dissemination. Services like presence and availability serve to publish, as widely as possible, the state of the principal on whose behalf these services run. An analogue to the publish concept has been missing in the PSTN.

The work described in this book has capitalized on the service architectures of the respective networks to allow individual service states to cross network boundaries. This enables the PSTN to avail itself of Internet-style services and also enables the Internet to leverage services already present in the PSTN. We have described architectures to render such crossover services possible.

We started in Chapter 4 by making a case for the use of the Session Initiation Protocol (SIP) as an underlying technology as well as the target protocol in our domain. It possesses all the properties we require in implementing our service-oriented architecture. In Chapter 6, we further validated the protocol as a distributed middleware in our domain.

Our work in Chapter 5 provided the call model mapping with state sharing (CMM/SS) technique, whereby Internet telephony endpoints can transparently leverage a subset of deployed services in the PSTN. Neither of the entities participating in the service is aware of such service sharing. The Internet telephony endpoint simply assumes that it is receiving services from the nearest intermediary (a proxy or a Back-to-Back User Agent (B2BUA)), and the PSTN service execution platform assumes that it is conversing with a traditional switch. As we point out in the chapter, not all PSTN services can be executed as transparently. The ones that cannot participate using our technique are inhibited more by the philosophical differences between the networks than by a deficiency in the CMM/SS technique.

Chapter 6 proposed an ontology to enable discrete events occurring in one network — the PSTN — to be transported to another network — the Internet — such that service execution could occur in the Internet. This work corrected a deficiency that we saw in the PSTN: although the PSTN has access to all types of interesting information, it has not been possible to disseminate this in a widespread manner. Doing so has the potential to create new services operating in synergy across both networks. We demonstrated how Internet-style services such as presence and availability are just as applicable to PSTN endpoints.

Chapter 7 built on the foundation of its predecessor to apply the ontology to the field of pervasive computing. We constructed a smart space in the telecommunications domain that makes it possible to use the information disseminated by the PSTN in creating new services. Once the information is easily captured and transported out of the PSTN, it is possible to create additional services beyond the presence, availability, and instant messaging. Service-oriented computing challenges researchers to find answers for the problems of service composition, service behavior in unfriendly environments, and providing trust and security in a hostile ecosystem. We have addressed these issues for services in the telecommunications domain. Service composition is aided by a common ontology. Trust and security aspects can be mitigated by judicious use of existing components and technologies, and the notion of self-tuning is one answer to service behavior in an unfriendly environment. Certainly, self-tuning as demonstrated by the migration of a Short Message Service (SMS) into an instant message (IM) for delivery on the Internet is one example in which a service (SMS) is applicable to an environment (Internet) it was not designed for.

This work has demonstrated the evolving nature of the PSTN; traditionally, the PSTN has been viewed as a static network when compared to Internet telephony. Researchers have questioned the usefulness and utility of the rather complex PSTN/Intelligent Network (IN) call model in the face of the more nimble Internet telephony protocol state machines [ACK99, CHA01]. Our work on call models has demonstrated two aspects that contribute to the collective knowledge in this field: First, the paucity of states in a call model is not indicative of the type of services that can be provided through it. Certainly the SIP state machine has a fewer number of states than the PSTN/IN call model, but the richness of the events in it make it possible to devise new techniques that allow the SIP state machine to access services that were written for a call model with more states and transitions. Second, the PSTN/IN call model is still a useful and valid model and can, in fact, contribute to Internet-style services such as presence, availability, and instant messaging. It has the potential to grow and even adapt to the new environment.

Having the PSTN publish events occurring in it for service execution on the Internet has effectively addressed the problem of the lack of information dissemination in the PSTN. The PSTN is a virtual storehouse of events, and until now, there was not a general-purpose framework to allow it to export events out of the network. Our work has proposed such a framework and has demonstrated the potential of realizing the services we outlined in Chapters 6 and 7.

Finally, the notion of services in computing is moving toward the Web services model [LEA04] consisting of the Service Object Access Protocol (SOAP) and the Universal Description, Discovery, and Integration (UDDI) [OAS02] model. UDDI is the name of a group of Web-based registries that expose information about a business or other entity and its technical interfaces (or application programming interfaces (APIs)). Such technologies make extensive use of the eXtensible Markup Language (XML) to represent the semantic and syntactical information exchange between communicating peers. We believe that our work, which uses SIP and XML documents to execute discrete services, is an early effort of applying the computing discipline's Web service infrastructure to the Internet telephony service infrastructure.

Appendix A

Network Working Group
Request for Comments: 3910
Category: Standards Track

V.K. Gurbani
A. Brusilovsky
I. Faynberg
Lucent Technologies, Inc.
J. Gato
Vodafone Espana
H. Lu
Bell Labs/Lucent Technologies, Inc.
M. Unmehopa
Lucent Technologies, Inc.
October 2004

The SPIRITS (Services in PSTN Requesting Internet Services) Protocol

Status of this Memo

This document specifies an Internet standards track protocol for the Internet community, and requests discussion and suggestions for improvements. Please refer to the current edition of the "Internet Official Protocol

Standards" (STD 1) for the standardization state and status of this protocol. Distribution of this memo is unlimited.

Copyright Notice

Abstract

This document describes the Services in PSTN (Public Switched Telephone Network) requesting Internet Services (SPIRITS) protocol. The purpose of the SPIRITS protocol is to support services that originate in the cellular or wireline PSTN and necessitate interactions between the PSTN and the Internet. On the PSTN side, the SPIRITS services are most often initiated from the Intelligent Network (IN) entities. Internet Call Waiting and Internet Caller-ID Delivery are examples of SPIRITS services, as are location-based services on the cellular network. The protocol defines the building blocks from which many other services can be built.

TABLE OF CONTENTS

1. Introduction

SPIRITS (Services in the PSTN Requesting Internet Services) is an IETF architecture and an associated protocol that enables call processing elements in the telephone network to make service requests that are then processed on Internet hosted servers. The term Public Switched Telephone Network (PSTN) is used here to include the wireline circuit-switched network, as well as the wireless circuit-switched network.

The earlier IETF work on the PSTN/Internet Interworking (PINT) resulted in the protocol (RFC 2848) in support of the services initiated in the reverse direction — from the Internet to PSTN.

This document has been written in response to the SPIRITS WG chairs call for SPIRITS Protocol proposals. Among other contributions, this document is based on:

- Informational "Pre-SPIRITS implementations" [10]
- Informational "The SPIRITS Architecture" [1]
- Informational "SPIRITS Protocol Requirements" [4]

1.1. Conventions used in this document

The key words "MUST", "MUST NOT", "REQUIRED", "SHALL", "SHALL NOT", "SHOULD", "SHOULD NOT", "RECOMMENDED", "MAY", and "OPTIONAL" in this document are to be interpreted as described in BCP 14, RFC 2119 [2].

2. Overview of operations

The purpose of the SPIRITS protocol is to enable the execution of services in the Internet based on certain events occurring in the PSTN. The term PSTN is used here to include all manner of switching; i.e. wireline circuit-switched network, as well as the wireless circuit-switched network.

In general terms, an Internet host is interested in getting notifications of certain events occurring in the PSTN. When the event of interest occurs, the PSTN notifies the Internet host. The Internet host can execute appropriate services based on these notifications.

Figure 1 contains the SPIRITS events hierarchy, including their subdivision in two discrete classes for service execution: events related to the setup, teardown and maintenance of a call and events un-related to call setup, teardown or maintenance. Example of the latter class of events are geolocation mobility events that are tracked by the cellular PSTN. SPIRITS will specify the framework to provide services for both of these types of events.

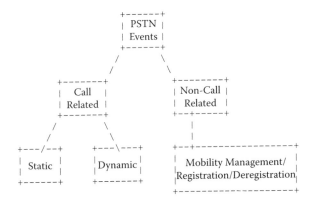

Figure 1   The SPIRITS Hierarchy.

Call-related events, its further subdivisions, and how they enable services in the Internet is contained in Section 5. Services enabled from events not related to call setup, teardown, or maintenance are covered in detail in Section 6.

For reference, the SPIRITS architecture from [1] is reproduced below. This document is focused on interfaces B and C only. Interface D is a matter of local policy; the PSTN operator may have a functional interface between the SPIRITS client or a message passing interface. This document does not discuss interface D in any detail.

The PSTN today supports service models such as the Intelligent Network (IN), whereby some features are executed locally on switching elements called Service Switching Points (SSPs). Other features are executed on service elements called Service Control Points (SCPs). The SPIRITS architecture [1] permits these SCP elements to act as intelligent entities to leverage and use Internet hosts and capabilities to further enhance the telephone end-user's experience.

The protocol used on interfaces B and C consists of the SPIRITS protocol, and is based on SIP and SIP event notification [3]. The requirements of a SPIRITS protocol and the choice of using SIP as an enabler are detailed in [4].

The SPIRITS protocol is a set of two "event packages" [3]. It contains the procedural rules and semantic context that must be applied to these rules for processing SIP transactions. The SPIRITS protocol has to carry subscriptions for events from the SPIRITS server to the SPIRITS client and notifications of these events from the SPIRITS client to the SPIRITS server. Extensible Markup Language (XML) [12] is used to codify the subscriptions and notifications.

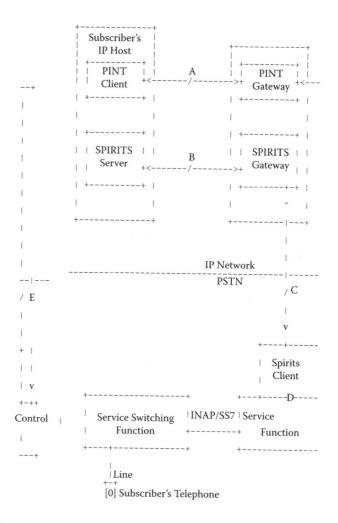

Figure 2 The SPIRITS Architecture.

Note: The interfaces A–E are described in detail in the SPIRITS Architecture document [1].

Finally, in the context of ensuing discussion, the terms "SPIRITSserver" and "SPIRITS client" are somewhat confusing since the roles appear reversed; to wit, the "SPIRITS server" issues a subscription which is accepted by a "SPIRITS client". To mitigate such ambiguity, from now on, we will refer to the "SPIRITS server" as a "SPIRITS subscriber" and to the "SPIRITS client" as a "SPIRITS notifier". This convention adheres to the nomenclature outlined in [3]; the SPIRITS server in Figure 2 is a subscriber (issues subscriptions to events), and the SPIRITS client in Figure 2 is a notifier (issues notifications whenever the event of interest occurs).

2.1. Terminology

For ease of reference, we provide a terminology of the SPIRITS actors discussed in the preceding above:

> Service Control Function (SCF): A PSTN entity that executes service logic. It provides capabilities to influence the call processing occurring in the Service Switching Function (SSF). For more information on how a SCF participates in the SPIRITS architecture, please see Sections 5 and 5.1.
> SPIRITS client: see SPIRITS notifier.
> SPIRITS server: see SPIRITS subscriber.
> SPIRITS notifier: A User Agent (UA) in the PSTN that accepts subscriptions from SPIRITS subscribers. These subscriptions contain events that the SPIRITS subscribers are interested in receiving a notification for. The SPIRITS notifier interfaces with the Service Control Function such that when the said event occurs, a notification will be sent to the relevant SPIRITS subscriber.
> SPIRITS subscriber: A UA in the Internet that issues a subscription containing events in the PSTN that it is interested in receiving a notification for.

3. Using XML for subscription and notification

The SPIRITS protocol requirements mandate that "SPIRITS-related parameters be carried in a manner consistent with SIP practices" (RFC3298:Section 3). SIP already provides payload description capabilities through the use of headers (Content-Type, Content-Length). This document defines a new MIME type — "application/spirits-event+xml" — and registers it with IANA (Section 7). This MIME type MUST be present in the "Content-Type" header of SPIRITS requests and responses, and it describes an XML document that contains SPIRITS-related information.

This document defines a base XML schema for subscriptions to PSTN events. The list of events that can be subscribed to is defined in the SPIRITS protocol requirements document [4] and this document provides an XML schema for it. All SPIRITS subscribers (any SPIRITS entity capable of issuing a SUBSCRIBE, REGISTER, or INVITE request) MUST support this schema. All SPIRITS notifiers (any SPIRITS entity capable of receiving and processing a SUBSCRIBE, REGISTER, or INVITE request) MUST support this schema. The schema is defined in Section 9.

> The support for the SIP REGISTER request is included for PINT compatibility (RFC3298:Section 6).

The support for the SIP INVITE request is mandated because pre-existing SPIRITS implementations did not use the SIP event notification scheme. Instead, the initial PSTN detection point always arrived via the SIP INVITE request.

This document also defines a base XML schema for notifications of events (Section 9). All SPIRITS notifiers MUST generate XML documents that correspond to the base notification schema. All SPIRITS subscribers MUST support XML documents that correspond to this schema.

The set of events that can be subscribed to and the amount of notification that is returned by the PSTN entity may vary among different PSTN operators. Some PSTN operators may have a rich set of events that can be subscribed to, while others have only the primitive set of events outlined in the SPIRITS protocol requirements document [4]. This document defines a base XML schema (in Section 9) which MUST be used for the subscription and notification of the primitive set of events. In order to support a richer set of event subscription and notification, implementations MAY use additional XML namespaces corresponding to alternate schemas in a SPIRITS XML document. However, all implementations MUST support the base XML schema defined in Section 9 of this document. Use of the base schema ensures interoperability across implementations, and the inclusion of additional XML namespaces allows for customization.

A logical flow of the SPIRITS protocol is depicted below (note: this example shows a temporal flow; XML documents and related SPIRITS protocol syntax is specified in later sections of this document). In the flow below, S is the SPIRITS subscriber and N is the SPIRITS notifier. The SPIRIT Gateway is presumed to have a pure proxying functionality and thus is omitted for simplicity:

1. S->N Subscribe (events of interest in an XML document instance using base subscription schema)
2. N->S 200 OK (Subscribe)
3. N->S Notify
4. S->N 200 OK (Notify communicating current resource state)
5. ...
6. N->S Notify (Notify communicating change in resource state; payload is an XML document instance using XML extensions to the base notification schema)
7. S->N 200 OK (Notify)

In line 1, the SPIRITS subscriber subscribes to certain events using an XML document based on the base schema defined in this document. In line 6, the SPIRITS notifier notifies the SPIRITS subscriber of the occurrence of the event using extensions to the base notification schema. Note that this document defines a base schema for event notification as well; the SPIRITS notifier could have availed itself of these. Instead, it chooses to pass to the SPIRITS subscriber an XML document composed of extensions to the base notification schema. The SPIRITS subscriber, if it understands the extensions, can interpret the XML document accordingly. However, in the event that the SPIRITS subscriber is not programmed to understand the extensions, it MUST search the XML document for the mandatory elements. These elements MUST be present in all notification schemas and are detailed in Section 9.

4. XML format definition

This section defines the XML-encoded SPIRITS payload format. Such a payload is a well formed XML document and is produced by SPIRITS notifiers and SPIRITS subscribers.

The namespace URI for elements defined in this document is a Uniform Resource Name (URN) [14], using the namespace identifier 'ietf' defined in [15] and extended by [16]:

urn:ietf:params:xml:ns:spirits-1.0

SPIRITS XML documents may have a default namespace, or they may be associated with a namespace prefix following the convention established in XML namespaces [17]. Regardless, the elements and attributes of SPIRITS XML documents MUST conform to the SPIRITS XML schema specified in Section 9.

The <spirits-event> element

The root of a SPIRITS XML document (characterized by a Content-Type header of "application/spirits-event+xml">) is the <spirits-event> element. This element MUST contain a namespace declaration ('xmlns') to indicate the namespace on which the XML document is based. XML documents compliant to the SPIRITS protocol MUST contain the URN "urn:ietf:params:xml:ns:spirits-1.0" in the namespace declaration. Other namespaces may be specified as needed.

<spirits-event> element MUST contain at least one <Event> element, and MAY contain more than one.

The <Event> element

The <Event> element contains three attributes, two of which are mandatory. The first mandatory attribute is a 'type' attribute whose value is either "INDPs" or "userprof".

These types correspond, respectively, to call-related events described in Section 5 and non-call related events described in Section 6.

The second mandatory attribute is a 'name' attribute. Values for this attribute MUST be limited to the SPIRITS mnemonics defined in Section 5.2.1, Section 5.2.2, and Section 6.1.

The third attribute, which is optional, is a 'mode' attribute. The value of 'mode' is either "N" or "R", corresponding respectively to (N)otification or (R)equest (RFC3298:Section 4). The default value of this attribute is "N".

If the 'type' attribute of the <Event> element is "INDPs", then it MUST contain at least one or more of the following elements (unknown elements MAY be ignored): <CallingPartyNumber>, <CalledPartyNumber>, <DialledDigits>, or <Cause>. These elements are defined in Section 5.2; they MUST not contain any attributes and MUST not be used further as parent elements. These elements contain a string value as described in Section 5.2.1 and 5.2.2.

If the 'type' attribute of the <Event> element is "userprof", then it MUST contain a <CalledPartyNumber> element and it MAY contain a <Cell-ID> element. None of these elements contain any attributes and neither must be used further as a parent element. These elements contain a string value as described in Section 6.1. All other elements MAY be ignored if not understood.

A SPIRITS-compliant XML document using the XML namespace defined in this document might look like the following example:

```
<?xml version="1.0" encoding="UTF-8"?>
<spirits-event xmlns="urn:ietf:params:xml:ns:spirits-1.0">
  <Event type="INDPs" name="OD" mode="N">
    <CallingPartyNumber>5551212</CallingPartyNumber>
  </Event>
```

```
    <Event type="INDPs" name="OAB" mode="N">
        <CallingPartyNumber>5551212</CallingPartyNumber>
    </Event>
</spirits-event>
```

5. Call-related events

For readers who may not be familiar with the service execution aspects of PSTN/IN, we provide a brief tutorial next. Interested readers are urged to consult [19] for a detailed treatment of this subject.

Services in the PSTN/IN are executed based on a call model. A call model is a finite state machine used in SSPs and other call processing elements that accurately and concisely reflects the current state of a call at any given point in time. Call models consist of states called PICs (Points In Call) and transitions between states. Inter-state transitions pass through elements called Detection Points or DPs. DPs house one or more triggers. Every trigger has a firing criteria associated with it. When a trigger is armed (made active), and its associated firing criteria are satisfied, it fires. The particulars of firing criteria may vary based on the call model being supported.

When a trigger fires, a message is formatted with call state information and transmitted by the SSP to the SCP. The SCP then reads this call related data and generates a response which the SSP then uses in further call processing.

Detection Points are of two types: TDPs (or Trigger Detection Points), and EDPs (or Event Detection Points). TDPs are provisioned with statically armed triggers (armed through Service Management Tools). EDPs are dynamically armed triggers (armed by the SCP as call processing proceeds). DPs may also be classified as "Request" or "Notification" DPs. Thus, one can have TDP-R's, TDP-N's, EDP-R's and EDP-N's.

The "-R" type of DPs require the SSP to suspend call processing when communication with the SCP is initiated. Call processing resumes when a response is received. The "-N" type of DPs enable the SSP to continue with call processing when the trigger fires, after it sends out the message to the SCP, notifying it that a certain event has occurred.

Call models typically support different types of detection points. Note that while INAP and the IN Capability Set (CS)-2 [7] call model are used in this document as examples, and for ease of explanation, other call models possess similar properties. For example, the Wireless Intelligent Network (WIN) call model also supports the dynamic arming of triggers. Thus, the essence of this discussion applies not just to the wireline domain, but applies equally well to the wireless domain as well.

When the SCP receives the INAP formatted message from the SSP, if the SCP supports the SPIRITS architecture, it can encode the INAP message contents into a SPIRITS protocol message which is then transmitted to SPIRITS-capable elements in the IP network. Similarly, when it receives responses back from said SPIRITS capable elements, it can reformat the response content into the INAP format and forward these messages back to SSPs. Thus the process of inter-conversion and/or encoding between the INAP parameters and the SPIRITS protocol is of primary interest.

An SCP is a physical manifestation of the Service Control Function. An SSP is a physical manifestation of the Service Switching Function (and the Call Control Function). To support uniformity of nomenclature between the various SPIRITS drafts, we shall use the terms SCP and SCF, and SSP and SSF interchangeably in this document.

5.1. IN-specific requirements

Section 4 of [4] outlines the IN-related requirements on the SPIRITS protocol. The SUBSCRIBE request arriving at the SPIRITS notifier MUST contain the events to be monitored (in the form of a DP list), the mode (request or a notification, the difference being that for a request, the SPIRITS subscriber can influence subsequent call processing and for a notification, no further influence is needed), and any DP-related parameters.

Section 4 of [4] also enumerates a list of Capability Set 3 (CS-3) DPs for SPIRITS services. It is a requirement (RFC3298:Section 4) that the SPIRITS protocol specify the relevant parameters of the DPs. These DPs and their relevant parameters to be carried in a SUBSCRIBE request are codified in an XML schema. All SPIRITS subscribers MUST understand this schema for subscribing to the DPs in the PSTN. The schema is defined in Section 9.

When a DP fires, a notification — using a SIP NOTIFY request — is transmitted from the SPIRITS notifier to the SPIRITS subscriber. The NOTIFY request contains an XML document which describes the DP that fired and any relevant parameters. The DPs and their relevant parameters to be carried in a NOTIFY request are codified in an XML schema. All SPIRITS notifiers MUST understand this schema; this schema MAY be extended. The schema is defined in Section 9.

In addition, Appendices A and B of [6] contain a select subset of CS-2 DPs that may be of interest to the reader. However, this document will only refer to CS-3 DPs outlined in [4].

5.2. Detection points and required parameters

The IN CS-3 DPs envisioned for SPIRITS services (RFC3298:Section 4) are described next. IN DPs are characterized by many parameters, however, not all such parameters are required — or even needed — by SPIRITS. This section, thus, serves to list the mandatory parameters for each DP that MUST be specified in subscriptions and notifications. Implementations can specify additional parameters as XML extensions associated with a private (or public and standardized) namespace.

The exhaustive list of IN CS-3 DPs and their parameters can be found in reference [13].

Each DP is given a SPIRITS-specific mnemonic for use in the subscriptions and notifications.

5.2.1.   Originating-side DPs

Origination Attempt Authorized
   SPIRITS mnemonic: OAA
   Mandatory parameter in SUBSCRIBE: CallingPartyNumber
   Mandatory parameters in NOTIFY: CallingPartyNumber,
   CalledPartyNumber
      CallingPartyNumber: A string used to identify the calling
      party for the call. The actual length and encoding of this
      parameter depend on the particulars of the dialing plan used.
      CalledPartyNumber: A string containing the number (e.g.,
      called directory number) used to identify the called party.
      The actual length and encoding of this parameter depend
      on the particulars of the dialing plan used.
Collected Information
   SPIRITS mnemonic: OCI
   Mandatory parameter in SUBSCRIBE: CallingPartyNumber
   Mandatory parameters in NOTIFY: CallingPartyNumber,
   DialledDigits
      DialledDigits: This parameter contains non-translated address
      information collected/received from the originating
      user/line/trunk
Analyzed Information
   SPIRITS mnemonic: OAI
   Mandatory parameter in SUBSCRIBE: CallingPartyNumber
   Mandatory parameters in NOTIFY: CallingPartyNumber,
   DialledDigits
Origination Answer
   SPIRITS mnemonic: OA
   Mandatory parameter in SUBSCRIBE: CallingPartyNumber
   Mandatory parameters in NOTIFY: CallingPartyNumber,
   CalledPartyNumber

Origination Term Seized
SPIRITS mnemonic: OTS
Mandatory parameter in SUBSCRIBE: CallingPartyNumber
Mandatory parameter in NOTIFY: CallingPartyNumber,
CalledPartyNumber
Origination No Answer
SPIRITS mnemonic: ONA
Mandatory parameter in SUBSCRIBE: CallingPartyNumber
Mandatory parameter in NOTIFY: CallingPartyNumber,
CalledPartyNumber
Origination Called Party Busy
SPIRITS mnemonic: OCPB
Mandatory parameter in SUBSCRIBE: CallingPartyNumber
Mandatory parameters in NOTIFY: CallingPartyNumber,
CalledPartyNumber
Route Select Failure
SPIRITS mnemonic: ORSF
Mandatory parameter in SUBSCRIBE: CallingPartyNumber
Mandatory parameter in NOTIFY: CallingPartyNumber,
CalledPartyNumber
Origination Mid Call
SPIRITS mnemonic: OMC
Mandatory parameter in SUBSCRIBE: CallingPartyNumber
Mandatory parameter in NOTIFY: CallingPartyNumber
Origination Abandon
SPIRITS mnemonic: OAB
Mandatory parameter in SUBSCRIBE: CallingPartyNumber
Mandatory parameter in NOTIFY: CallingPartyNumber
Origination Disconnect
SPIRITS mnemonic: OD
Mandatory parameter in SUBSCRIBE: CallingPartyNumber
Mandatory parameter in NOTIFY: CallingPartyNumber,
CalledPartyNumber

5.2.2   Terminating-side DPs

Termination Answer
SPIRITS mnemonic: TA
Mandatory parameter in SUBSCRIBE: CalledPartyNumber
Mandatory parameters in NOTIFY: CallingPartyNumber,
CalledPartyNumber

Termination No Answer
    SPIRITS mnemonic: TNA
    Mandatory parameter in SUBSCRIBE: CalledPartyNumber
    Mandatory parameters in NOTIFY: CallingPartyNumber,
    CalledPartyNumber
Termination Mid-Call
    SPIRITS mnemonic: TMC
    Mandatory parameter in SUBSCRIBE: CalledPartyNumber
    Mandatory parameter in NOTIFY: CalledPartyNumber
Termination Abandon
    SPIRITS mnemonic: TAB
    Mandatory parameter in SUBSCRIBE: CalledPartyNumber
    Mandatory parameter in NOTIFY: CalledPartyNumber
Termination Disconnect
    SPIRITS mnemonic: TD
    Mandatory parameter in SUBSCRIBE: CalledPartyNumber
    Mandatory parameters in NOTIFY: CalledPartyNumber,
    CallingPartyNumber
Termination Attempt Authorized
    SPIRITS mnemonic: TAA
    Mandatory parameter in SUBSCRIBE: CalledPartyNumber
    Mandatory parameters in NOTIFY: CalledPartyNumber,
    CallingPartyNumber
Termination Facility Selected and Available
    SPIRITS mnemonic: TFSA
    Mandatory parameter in SUBSCRIBE: CalledPartyNumber
    Mandatory parameter in NOTIFY: CalledPartyNumber
Termination Busy
    SPIRITS mnemonic: TB
    Mandatory parameter in SUBSCRIBE: CalledPartyNumber
    Mandatory parameters in NOTIFY: CalledPartyNumber,
    CallingPartyNumber, Cause
        Cause: This parameter contains a string value of either "Busy"
        or "Unreachable". The difference between these is translated
        as a requirement (RFC3298:Section 5) to aid in the SPIRITS
        subscriber in determining if the called party is indeed busy
        (engaged), or if the called party is unavailable (as it would
        be if it were on the cellular PSTN and the mobile subscriber
        was not registered with the network).

5.3. Services through dynamic DPs

Triggers in the PSTN can be armed dynamically, often outside the context of a call. The SIP event notification mechanism [3] is, therefore, a convenient means to exploit in those cases where triggers housed in EDPs fire (see section 3 of [4]). Note that [4] uses the term "persistent" to refer to call-related DP arming and associated interactions.

The SIP Events Package enables IP endpoints (or hosts) to subscribe to and receive subsequent notification of events occurring in the PSTN. With reference to Figure 2, this includes communication on the interfaces marked "B" and "C".

5.3.1. Normative usage

A subscriber will issue a SUBSCRIBE request which identifies a set of events (DPs) it is interested in getting the notification of. This set MUST contain at least one DP, it MAY contain more than one. The SUBSCRIBE request is routed to the notifier, where it is accepted, pending a successful authentication.

When any of the DPs identified in the set of events fires, the notifier will format a NOTIFY request and direct it towards the subscriber. The NOTIFY request will contain information pertinent to the event that was triggered. The un-encountered DPs MUST be subsequently dis-armed by the SPIRITS notifier and/or the SCF.

The dialog established by the SUBSCRIBE terminates when the event of interest occurs and this notification is passed to the subscriber through a NOTIFY request. If the subscriber is interested in the future occurrence of the same event, it MUST issue a new SUBSCRIBE request, establishing a new dialog.

When the subscriber receives a NOTIFY request, it can subsequently choose to act in a manner appropriate to the notification. The remaining sections fill in the specific package responsibilities raised in RFC3265 [3], Section 4.4.

5.3.2. Event package name

This document defines two event packages; the first of these is defined in this section and is called "spirits-INDPs". This package MUST be used for events corresponding to IN detection points in the cellular or wireline PSTN. All entities that implement the SPIRITS protocol and support IN detection points MUST set the "Event" request header [3] to "spirits-INDPs."

The "Allow-Events" general header [3] MUST include the token "spirits-INDPs" if the entity implements the SPIRITS protocol and supports IN detection points.

> Event: spirits-INDPs
> Allow-Events: spirits-INDPs

The second event package is defined and discussed in Section 6.

5.3.3. Event package parameters

The "spirits-INDPs" event package does not support any additional parameters to the Event header.

5.3.4. SUBSCRIBE bodies

SUBSCRIBE requests that serve to terminate the subscription MAY contain an empty body; however, SUBSCRIBE requests that establish a dialog MUST contain a body which encodes three pieces of information:

(1) The set of events (DPs) that is being subscribed to. A subscriber MAY subscribe to multiple DPs in one SUBSCRIBE request, or MAY issue a different SUBSCRIBE request for each DP it is interested in receiving a notification for. The protocol allows for both forms of representation, however, it recommends the former manner of subscribing to DPs if the service depends on any of the DPs being triggered.

(2) Because of the requirement [4] that IN be informed whether the detection point is set as the request or notification, all events in the "spirits-INDPs" package (but not in the "spirits-user-prof" package) are required to provide a "mode" parameter, whose values are "R" (for Request) and "N" for notification.

(3) A list of the values of the parameters associated with the event detection point (Note: the term "event" here refers to the IN usage — a dynamically armed DP is called an Event Detection Point). Please see Section 5.2.1 and Section 5.2.2 for a list of parameters associated with each DP.

The default body type for SUBSCRIBEs in SPIRITS is denoted by the MIME type "application/spirits-event+xml". The "Accept" header, if present, MUST include this MIME type.

5.3.5. Subscription duration

For package "spirits-INDPs", the purpose of the SUBSCRIBE request is to arm the DP, since as far as IN is concerned, being armed is the first essential pre-requisite. A DP maybe armed either statically (for instance, through service provisioning), or dynamically (by the SCF). A statically armed DP remains armed until it is disarmed proactively. A dynamically armed DP remains armed for the duration of a call (or more appropriately, no longer than the duration of a particular SSF-SCF relationship). Dynamically armed DPs are automatically disarmed when the event of interest occurs in the notifier. It is up to the subscriber to re-arm the DPs within the context of a call, if it so desires.

Statically armed DPs are considered outside the scope of the SPIRITS protocol requirements [4] and thus will not be considered any further.

5.3.6. NOTIFY bodies

Bodies in NOTIFY requests for the "spirits-INDPs" package are optional. If present, they MUST be of the MIME type "application/spirits-event+xml". The body in a NOTIFY request encapsulates the following pieces of information which can be used by the subscriber:

(1) The event that resulted in the NOTIFY being generated (typically, but not always, this will be the same event present in the corresponding SUBSCRIBE request).

(2) The "mode" parameter; it is simply reflected back from the corresponding SUBSCRIBE request.

(3) A list of values of the parameters associated with the event that the NOTIFY is being generated for. Depending on the actual event, the list of the parameters will vary.

If the subscriber armed multiple DPs as part of a single SUBSCRIBE request, all the un-encountered DPs that were part of the same SUBSCRIBE dialog MUST be dis-armed by the SPIRITS notifier and/or the SCF/SCP.

5.3.7. Notifier processing of SUBSCRIBE requests

When the notifier receives a SUBSCRIBE request, it MUST authenticate the request and ensure that the subscriber is authorized to access the resource being subscribed to, in this case, PSTN/IN events on a certain PSTN line.

Once the SUBSCRIBE request has been authenticated and authorized, the notifier interfaces with the SCF over interface D to arm the detection points corresponding to the PSTN line contained in the SUBSCRIBE body. The particulars about interface D is out of scope for this document; here we will simply assume that the notifier can affect the arming (and disarming) of triggers in the PSTN through interface D.

5.3.8. Notifier generation of NOTIFY requests

If the notifier expects the arming of triggers to take more than 200 ms, it MUST send a 202 response to the SUBSCRIBE request immediately, accepting the subscription. It should then send a NOTIFY request with an empty body. This NOTIFY request MUST have a "Subscription-State" header with a value of "pending".

> This immediate NOTIFY with an empty body is needed since the resource identified in the SUBSCRIBE request does not have as yet a meaningful state.

Once the notifier has successfully interfaced with the SCF, it MUST send a subsequent NOTIFY request with an empty body and a "Subscription-State" header with a value of "active."

When the event of interest identified in the SUBSCRIBE request occurs, the notifier sends out a new NOTIFY request which MUST contain a body (see Section 5.3.6). The NOTIFY request MUST have a "Subscription-State" header and its value MUST be set to "terminated" with a reason parameter of "fired".

5.3.9. Subscriber processing of NOTIFY requests

The exact steps executed at the subscriber when it gets a NOTIFY request will depend on the service being implemented. As a generality, the UA associated with the subscriber should somehow impart this information to the user by visual or auditory means, if at all possible.

If the NOTIFY request contained a "Subscription-State" header with a value of "terminated" and a reason parameter of "fired", the UA associated with the subscriber MAY initiate a new subscription for the event that was just reported through the NOTIFY request.

> Whether or not to initiate a new subscription when an existing one expires is up to the context of the service that is being implemented. For instance, a user may configure her UA to always re-subscribe to the same event when it fires, but this is not necessarily the normative case.

5.3.10. Handling of forked requests

Forking of SUBSCRIBE requests is prohibited. Since the SUBSCRIBE request is targeted towards the PSTN, highly irregular behaviors occur if the request is allowed to fork. The normal SIP DNS lookup and routing rules [11] should result in a target set with exactly one element: the notifier.

5.3.11. Rate of notifications

For reasons of security more than network traffic, it is RECOMMENDED that the notifier issue two or, at most three NOTIFY requests for a subscription. If the subscription was accepted with a 202 response, a NOTIFY will be sent immediately towards the subscriber. This NOTIFY serves to inform the subscriber that the request has been accepted and is being acted on.

Once the resource (detection points) identified in the SUBSCRIBE request have been initialized, the notifier MUST send a second NOTIFY request. This request contains the base state of the resource.

When an event of interest occurs which leads to the firing of the trigger associated with the detection points identified in the SUBSCRIBE request, a final NOTIFY is sent to the subscriber. This NOTIFY request contains more information about the event of interest.

If the subscription was accepted with a 200 response, the notifier simply sends two NOTIFY requests: one containing the base state of the resource, and the other containing information that lead to the firing of the detection point.

5.3.12. State agents

State agents are not used in SPIRITS.

5.3.13. Examples

This section contains example call flows for a SPIRITS service called Internet Caller-ID Delivery (ICID). One of the benchmark SPIRITS service, as described in section 2.2 of [1] is Internet Caller-ID delivery:

> This service allows the subscriber to see the caller's number or name or both while being connected to the Internet. If the subscriber has only one telephone line and is using the very line for the Internet connection, the service is a subset of the ICW service and follows the relevant description in Section 2.1.

Otherwise, the subscriber's IP host serves as an auxiliary device of the telephone to which the call is first sent.

We present an example of a SPIRITS call flow to realize this service. Note that this is an example only, not a normative description of the Internet Caller-ID service.

Further text and details of SIP messages below refer to the call flow provided in Figure 3. Figure 3 depicts the 4 entities that are an integral part of any SPIRITS service (the headings of the entities refer to the names established in Figure 1 in [1]) — the SPIRITS subscriber, the SPIRITS notifier and the SCF. Note that the SPIRITS gateway is not included in this figure; logically, SPIRITS messages flow between the SPIRITS server and the SPIRITS client. A gateway, if present, may act as a proxy.

This call flow depicts an overall operation of a "subscriber" successfully subscribing to the IN Termination_Attempt_Authorized DP (the "subscriber" is assumed to be a user, possibly at work, who is interested in knowing when he/she gets a phone call to his/her home phone number) — this interaction is captured in messages F1 through F8 in Figure 3. The user sends (F1) a SIP SUBSCRIBE request identifying the DP it is interested in along with zero or more parameters relevant to that DP (in this example, the Termination_Attempt_DP will be employed). The SPIRITS notifier in

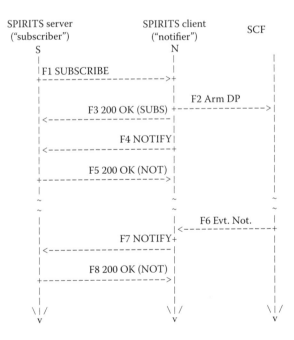

Figure 3   Sample call flow

turns interacts with the SCF to arm the Termination_Attempt_DP for the service (F2). An immediate NOTIFY with the current state information is send to the subscriber (F4, F5).

At some point after the above sequence of events has transpired, the PSTN gets a call to the users phone. The SSF informs the SCF of this event when it encounters an armed Termination_Attempt_DP (not shown in Figure 3). The SCF informs the SPIRITS notifier of this event (F6).

When the SPIRITS notifier receives this event, it forms a SIP NOTIFY request and directs it to the SPIRITS subscriber (F7). This NOTIFY will contain all the information elements necessary to identify the caller to the subscriber. The subscriber, upon receiving the notification (F8) may pop open a window with the date/time and the number of the caller.

The rest of this section contains the details of the SIP messages in Figure 3. The call flow details below assume that the SPIRITS gateway is, for the purpose of this example, a SIP proxy that serves as the default outbound proxy for the notifier and an ingress host of the myprovider.com domain for the subscriber. The subscriber and notifier may be in separate administrative domains.

F1: S->N

```
SUBSCRIBE sip:myprovider.com SIP/2.0
From: <sip:vkg@example.com>;tag=8177-afd-991
To: <sip:16302240216@myprovider.com>
CSeq: 18992 SUBSCRIBE
Call-ID: 3329as77@host.example.com
Contact: <sip:vkg@host.example.com>
Via: SIP/2.0/UDP host.example.com;branch=z9hG4bK776asdhds
Expires: 3600
Event: spirits-INDPs
Allow-Events: spirits-INDPs, spirits-user-prof
Accept: application/spirits-event+xml
Content-Type: application/spirits-event+xml
Content-Length: ...
```

```
<?xml version="1.0" encoding="UTF-8"?>
<spirits-event xmlns="urn:ietf:params:xml:ns:spirits-1.0">
  <Event type="INDPs" name="TAA" mode="N">
      <CalledPartyNumber>6302240216</CalledPartyNumber>
  </Event>
</spirits-event>
```

The subscriber forms a SIP SUBSCRIBE request which identifies the DP that it wants to subscribe to (in this case, the TAA DP) and the actual line it wants that DP armed for (in this case, the line associated with the phone number 6302240216). This request eventually arrives at the SIPRITS notifier, N, which authenticates it (not shown) and sends a successful response to the subscriber:

F3: N->S

```
SIP/2.0 200 OK
From: <sip:vkg@example.com>;tag=8177-afd-991
To: <sip:16302240216@myprovider.com>;tag=SPIRITS-TAA-
6302240216
CSeq: 18992 SUBSCRIBE
Call-ID: 3329as77@host.example.com
Contact: <sip:notifier.myprovider.com>
Via: SIP/2.0/UDP host.example.com;branch=z9hG4bK776asdhds
Expires: 3600
Accept: application/spirits-event+xml
Content-Length: 0
```

The notifier interacts with the SCF to arm the DP and also sends an immediate NOTIFY towards the subscriber informing the subscriber of the current state of the notification:

F4: N->S

```
NOTIFY sip:vkg@host.example.com SIP/2.0
From: <sip:16302240216@myprovider.com>;tag=SPIRITS-TAA-
6302240216
To: <sip:vkg@example.com>;tag=8177-afd-991
Via: SIP/2.0/UDP gateway.myprovider.com;branch=z9hG4bK-9$0-1
Via: SIP/2.0/UDP notifier.myprovider.com;branch=z9hG4bKqo--9
Call-ID: 3329as77@host.example.com
Contact: <sip:notifier.myprovider.com>
Subscription-State: active
CSeq: 3299 NOTIFY
Accept: application/spirits-event+xml
Content-Length: 0
```

F5: S->N

SIP/2.0 200 OK
From: <sip:16302240216@myprovider.com>;tag=SPIRITS-TAA-6302240216
To: <sip:vkg@example.com>;tag=8177-afd-991
Via: SIP/2.0/UDP gateway.myprovider.com;branch=z9hG4bK-9$0-1
Via: SIP/2.0/UDP notifier.myprovider.com;branch=z9hG4bKqo--9
Call-ID: 3329as77@host.example.com
Contact: <sip:vkg@host.example.com>
CSeq: 3299 NOTIFY
Accept: application/spirits-event+xml
Content-Length: 0

At some later point in time (before the subscription established in F1 expires at the notifier), a call arrives at the number identified in XML-encoded body of F1 — 6302240216. The SCF notifies the notifier (F6). Included in this notification is the relevant information from the PSTN, namely, the phone number of the party attempting to call 6302240216. The notifier uses this information to create a SIP NOTIFY request and sends it to the subscriber. The SIP NOTIFY request has a XML-encoded body with the relevant information from the PSTN:

F7: N->S

NOTIFY sip:vkg@host.example.com SIP/2.0
From: <sip:16302240216@myprovider.com>;tag=SPIRITS-TAA-6302240216
To: <sip:vkg@example.com>;tag=8177-afd-991
Via: SIP/2.0/UDP notifier.myprovider.com;branch=z9hG4bK9inn-=u7
Call-ID: 3329as77@host.example.com
Contact: <sip:notifier.myprovider.com>
CSeq: 3300 NOTIFY
Subscription-State: terminated;reason=fired
Accept: application/spirits-event+xml
Event: spirits-INDPs
Allow-Events: spirits-INDPs, spirits-user-prof
Content-Type: application/spirits-event+xml
Content-Length: …

```
<?xml version="1.0" encoding="UTF-8"?>
<spirits-event xmlns="urn:ietf:params:xml:ns:spirits-1.0">
  <Event type="INDPs" name="TAA" mode="N">
    <CalledPartyNumber>6302240216</CalledPartyNumber>
    <CallingPartyNumber>3125551212</CallingPartyNumber>
  </Event>
</spirits-event>
```

There are two important issues to note in the call flows for F7:

(1) The body of the NOTIFY request contains the information passed to the SPIRITS notifier from the SCF. In this particular example, this is the phone number of the party (3125551212) that attempted to call 6302240216.

(2) Since the notification occurred, the subscription established in F1 terminated (as evident by the Subscription-State header). The subscription terminated normally due to the DP associated with TAA firing (hence the reason code of "fired" in the Subscription-State header). If the subscriber wants to get notified of another attempt to call the number 6302240216, he/she should send a new SUBSCRIBE request to the notifier.

The subscriber can take any appropriate action upon the receipt of the NOTIFY in F7. A reasonable implementation may pop up a window populated with the information contained in the body of F12, along with a button asking the subscriber if they would like to re-subscribe to the same event. Alternatively, a re-subscription could be generated automatically by the subscriber's UA based on his/her preferences.

To complete the protocol, the subscriber also sends a 200 OK message towards the notifier:

F8: S->N

200 OK SIP/2.0
From: <sip:16302240216@myprovider.com>;tag=SPIRITS-TAA-6302240216
To: <sip:vkg@example.com>;tag=8177-afd-991
Via: SIP/2.0/UDP notifier.myprovider.com;z9hG4bK9inn-=u7
Call-ID: 3329as77@host.example.com
CSeq: 3300 NOTIFY
Content-Length: 0

5.3.14.   Use of URIs to retrieve state

The "spirits-INDPs" package MUST NOT use URIs to retrieve state. It is expected that most state information for this package is compact enough to fit in a SIP message. However, to err on the side of caution, implementations MUST follow the convention outlined in Section 18.1.1 of [5] and use a congestion controlled transport if the size of the request is within 200 bytes of the path MTU if known, or if the request size is larger than 1300 bytes and the path MTU is unknown.

5.4.   Services through static DPs

We mentioned in Section 5.1 that the first trigger that fires during call processing is typically a TDP since there isn't any pre-existing control relationship between the SSF and the SCF. Some Internet hosts may have expressed an interest in executing services based on TDPs (through an a-priori arrangement, which is not a part of this specification). Thus, the PSTN will notify such hosts. To do so, it will send a SIP request (typically an INVITE) towards the Internet host. The body of the SIP request MUST contain multi-part MIME with two MIME components: the first part corresponding to the normal payload, if any, of the request; and the second part will contain SPIRITS-specific information (e.g., the DP that fired). Responses to the INVITE request, or subsequent SUBSCRIBE messages from the Internet host to the PSTN within a current call context may result in EDPs being armed.

5.4.1.   Internet Call Waiting (ICW)

ICW as a benchmark SPIRITS service actually predates SPIRITS itself. Pre-SPIRITS implementations of ICW are detailed in [10]. However, as the document notes, while a diversity of implementations exists, these implementations are not interoperable. At the time [10] was published, the industry did not have the depth of experience with SIP as is the case now. The use of SIP in [10] does not constitute normative usage of SIP as described in [5]; for instance, no mention is made of the SDP (if any) in the initial INVITE (especially since this pertains to "accept the call using VoIP" case). Thus this section serves to provide a normative description of ICW in SPIRITS. The description of ICW is deceptively simple: it is a service most useful for single line phone subscribers that use the line to establish an Internet session. In a nutshell, the service enables subscriber engaged in an Internet dial-up session to

- ■ be notified of an incoming call to the very same telephone line that is being used for the Internet connection,

■ specify the desirable treatment of the call, and
■ have the call handled as specified.

5.4.2. Call disposition choices

Section 2 of [10] details the call disposition outcome of a ICW session. They are reproduced here as a numbered list for further discussion:

1. Accepting the call over the PSTN line, thus terminating the Internet (modem) connection
2. Accepting the call over the Internet using Voice over IP (VoIP)
3. Rejecting the call
4. Playing a pre-recorded message to the calling party and disconnecting the call
5. Forwarding the call to voice mail
6. Forwarding the call to another number
7. Rejecting (or Forwarding) on no Response — If the subscriber fails to respond within a certain period of time after the dialog box has been displayed, the incoming call can be either rejected or handled based on the treatment pre-defined by the subscriber.

It should be pointed out for the sake of completeness that ICW as a SPIRITS service is not possible without making the SCP aware of the fact that the subscriber line is being used for an Internet session. That awareness, however, is not a part of the ICW service, but solely a pre-requisite. One of the following three methods MUST be utilized to impart this information to the SCP:

A. ICW subscriber based method: the ICW client on the subscriber's PC notifies the SCP of the Internet session by issuing a SIP REGISTER request.
B. IN based method: SCP maintains a list of Internet Service Provider (ISP) access numbers for a geographical area; when one of these numbers is dialed and connected to, it (the SCP) assumes that the calling party is engaged in an Internet session.
C. Any combination of methods A and B.

ICW depends on a TDP to be provisioned in the SSP. When the said TDP is encountered, the SSP suspends processing of the call and sends a request to the SPIRITS-capable SCP. The SCP determines that the subscriber line is being used for an Internet session. It instructs the SPIRITS notifier on the SCP to create a SIP INVITE request and send it to the SPIRITS subscriber running on the subscriber's IP host.

The SPIRITS subscriber MUST return one of the possible call disposition outcomes catalogued in Section 5.4.2. Note that outcomes 1 and 4 through 7 can all be coalesced into one case, namely redirecting (using the SIP 3xx response code) the call to an alternative SIP URI. In case of 1, the URI of the redirected call MUST match the very same number being used by the customer to get online. Rejecting the call implies sending a non-2xx and non-3xx final response; the remaining outcomes result in the call being redirected to an alternate URI which provides the desired service (i.e., play a pre-recorded announcement, or record a voice message).

Further processing of a SPIRITS notifier when it receives a final response can be summarized by the following steps:

1. If the response is a 4xx, 5xx, or 6xx class of response, generate and transmit an ACK request and instruct the SSP to play a busy tone to the caller.
2. Else, for all 3xx responses, generate and transmit an ACK request, and compare the redirected URI to the subscriber's line number:
   2a. If the comparison indicates a match, instruct the SSP to hold onto the call for just enough time to allow the SPIRITS subscriber to disconnect the modem, thus freeing up the line; and then continue with normal call processing, which will result in the subscriber's phone to ring.
   2b. If the comparison fails, instruct the SSP to route the call to the redirected URI.
3. Else, for a 2xx response, follow the steps in section 5.4.3.

5.4.3. Accepting an ICW session using VoIP

One call handling option in ICW is to "accept an incoming call using VoIP". The SPIRITS notifier has no way of knowing a-priori if the subscriber (callee) will be choosing this option; nonetheless, it has to account for such a choice by adding a SDP in the body of the INVITE request. A possible way of accomplishing this is to have the SPIRITS notifier control a PSTN gateway and allocate appropriate resources on it. Once this is done, the SPIRITS notifier adds network information (IP address of the gateway and port numbers where media will be received) and codec information as the SDP portion of the body in the INVITE request. SPIRITS requires the DP information to be carried in the request body as well. To that extent, the SPIRITS notifier MUST also add the information associated with the TDP that triggered the service. Thus, the body of the INVITE MUST contain multi-part MIME, with two components.

The SPIRITS notifier transmits the INVITE request to the subscriber and now waits for a final response. Further processing when the SPIRITS subscriber returns a 200 OK MUST be handled as follows:

On the receipt of a 200 OK containing the SDP of the subscriber's UA, the SPIRITS notifier will instruct the SSP to terminate the call on a pre-allocated port on the gateway. This port MUST be correlated by the gateway to the SDP that was sent in the earlier INVITE.

The end result is that the caller and callee hold a voice session with part of the session occurring over VoIP.

6. Non-call related events

There are network events that are not related to setting up, maintaining, or tearing down voice calls. Such events occur on the cellular wireless network and can be used by SPIRITS to provide services. The SPIRITS protocol requirement explicitly includes the following events for which SPIRITS notification is needed (RFC3298:Section 5(b)):

1. Location update in the same Visitor Location Register (VLR) service area
2. Location update in another VLR service area
3. International Mobile Subscriber Identity (IMSI) attach
4. Mobile Subscriber (MS) initiated IMSI detach
5. Network initiated IMSI detach

6.1. Non-call events and their required parameters

Each of the five non-call related event is given a SPIRITS-specific mnemonic for use in subscriptions and notifications.

Location update in the same VLR area
SPIRITS mnemonic: LUSV
Mandatory parameter in SUBSCRIBE: CalledPartyNumber
Mandatory parameter in NOTIFY: CalledPartyNumber, Cell-ID
Cell-ID: A string used to identify the serving Cell-ID. The actual length and representation of this parameter depend on the particulars of the cellular provider's network.
Location update in different VLR area
SPIRITS mnemonic: LUDV
Mandatory parameter in SUBSCRIBE: CalledPartyNumber
Mandatory parameter in NOTIFY: CalledPartyNumber, Cell-ID
IMSI attach
SPIRITS mnemonic: REG

Mandatory parameter in SUBSCRIBE: CalledPartyNumber
Mandatory parameter in NOTIFY: CalledPartyNumber, Cell-ID
MS initiated IMSI detach
SPIRITS mnemonic: UNREGMS
Mandatory parameter in SUBSCRIBE: CalledPartyNumber
Mandatory parameter in NOTIFY: CalledPartyNumber
Network initiated IMSI detach
SPIRITS mnemonic: UNREGNTWK
Mandatory parameter in SUBSCRIBE: CalledPartyNumber
Mandatory parameter in NOTIFY: CalledPartyNumber

6.2. Normative usage

A subscriber will issue a SUBSCRIBE request which identifies a set of non-call related PSTN events it is interested in getting the notification of. This set MAY contain exactly one event, or it MAY contain multiple events. The SUBSCRIBE request is routed to the notifier where it is accepted, pending a successful authentication.

When any of the events identified in the set occurs, the notifier will format a NOTIFY request and direct it towards the subscriber. The NOTIFY request will contain information pertinent to the one of the event whose notification was requested.

The dialog established by the SUBSCRIBE persists until it expires normally, or is explicitly expired by the subscriber. This behavior is different than the behavior for subscriptions associated with the "spirits-INDPs" package. In the cellular network, the events subscribed for may occur at a far greater frequency than those compared to the wireline network (consider location updates as a cellular user moves around). Thus it is far more expedient to allow the subscription to expire normally.

When a subscriber receives a NOTIFY request, it can subsequently choose to act in a manner appropriate to the notification.

The remaining sections fill in the specific package responsibilities raised in RFC3265 [3], Section 4.4.

6.3. Event package name

This document defines two event packages; the first was defined in Section 5.3. The second package, defined in this section is called "spirits-user-prof". This package MUST be used for events corresponding to non-call related events in the cellular network. All entities that implement the SPIRITS protocol and support the non-call related events outlined in the SPIRITS protocol requirements (RFC3298:Section 5(b)) MUST set the "Event" header request header[3] to "spirits-user-prof." The "Allow-Events" general header [3] MUST include the token "spirits-user-prof" as well.

Example:

Event: spirits-user-prof
Allow-Events: spirits-user-prof, spirits-INDPs

6.4. Event package parameters

The "spirits-user-prof" event package does not support any additional parameters to the Event header.

6.5. SUBSCRIBE bodies

SUBSCRIBE requests that serve to terminate the subscriptions MAY contain an empty body; however, SUBSCRIBE requests that establish a dialog MUST contain a body which encodes two pieces of information:

(1) The set of events that is being subscribed to. A subscriber MAY subscribe to multiple events in one SUBSCRIBE request, or MAY issue a different SUBSCRIBE request for each event it is interested in receiving a notification for. The protocol allows for both forms of representation. However, note that if one SUBSCRIBE is used to subscribe to multiple events, then an expiry for the dialog associated with that subscription affects all such events.

(2) A list of values of the parameters associated with the event. Please see Section 6.1 for a list of parameters associated with each event.

The default body type for SUBSCRIBEs in SPIRITS is denoted by the MIME type "application/spirits-event+xml". The "Accept" header, if present, MUST include this MIME type.

6.6. Subscription duration

The duration of a dialog established by a SUBSCRIBE request is limited to the expiration time negotiated between the subscriber and notifier when the dialog was established. The subscriber MUST send a new SUBSCRIBE to refresh the dialog if it is interested in keeping it alive. A dialog can be terminated by sending a new SUBSCRIBE request with an "Expires" header value of 0, as outlined in [3].

6.7.   NOTIFY bodies

Bodies in NOTIFY requests for the "spirits-user-prof" package are optional. If present, they MUST be of the MIME type "application/spirits-event+xml". The body in a NOTIFY request encapsulates the following pieces of information which can be used by the subscriber:

> (1)   The event that resulted in the NOTIFY being generated (typically, but not always, this will be the same event present in the corresponding SUBSCRIBE request).
>
> (2)   A list of values of the parameters associated with the event that the NOTIFY is being generated for. Depending on the actual event, the list of the parameters will vary.

6.8.   Notifier processing of SUBSCRIBE requests

When the notifier receives a SUBSCRIBE request, it MUST authenticate the request and ensure that the subscriber is authorized to access the resource being subscribed to, in this case, non-call related cellular events for a mobile phone.

Once the SUBSCRIBE request has been authenticated and authorized, the notifier interfaces with the SCF over interface D to set marks in the HLR corresponding to the mobile phone number contained in the SUBSCRIBE body. The particulars of interface D are outside the scope of this document; here we simply assume that the notifier is able to set the appropriate marks in the HLR.

6.9.   Notifier generation of NOTIFY requests

If the notifier expects the setting of marks in the HLR to take more than 200 ms, it MUST send a 202 response to the SUBSCRIBE request immediately, accepting the subscription. It should then send a NOTIFY request with an empty body. This NOTIFY request MUST have a "Subscription-State" header with a value of "pending".

> This immediate NOTIFY with an empty body is needed since the resource identified in the SUBSCRIBE request does not have as yet a meaningful state.

Once the notifier has successfully interfaced with the SCF, it MUST send a subsequent NOTIFY request with an empty body and a "Subscription-State" header with a value of "active."

When the event of interest identified in the SUBSCRIBE request occurs, the notifier sends out a new NOTIFY request which MUST contain a body as described in Section 6.7.

6.10. Subscriber processing of NOTIFY requests

The exact steps executed at the subscriber when it receives a NOTIFY request depend on the nature of the service that is being implemented. As a generality, the UA associated with the subscriber should somehow impart this information to the user by visual or auditory means, if at all possible.

6.11. Handling of forked requests

Forking of SUBSCRIBE requests is prohibited. Since the SUBSCRIBE request is targeted towards the PSTN, highly irregular behaviors occur if the request is allowed to fork. The normal SIP DNS lookup and routing rules [11] should result in a target set with exactly one element: the notifier.

6.12. Rate of notifications

For reasons of congestion control, it is important that the rate of notifications not become excessive. For instance, if a subscriber subscribes to the location update event for a notifier moving through the cellular network at a high enough velocity, it is entirely conceivable that the notifier may generate many NOTIFY requests in a small time frame. Thus, within this package, the location update event needs an appropriate throttling mechanism.

Whenever a SPIRITS notifier sends a location update NOTIFY, it MUST start a timer (Tn) with a value of 15 seconds. If a subsequent location update NOTIFY request needs to be sent out before the timer has expired, it MUST be discarded. Any future location update NOTIFY requests MUST be transmitted only if Tn has expired (i.e. 15 seconds have passed since the last NOTIFY request was send out). If a location update NOTIFY is send out, Tn should be reset to go off again in 15 seconds.

6.13. State agents

State agents are not used in SPIRITS.

6.14. Examples

This section contains an example of a SPIRITS service that may be used to update the presence status of a mobile user. The call flow is depicted in Figure 4 below.

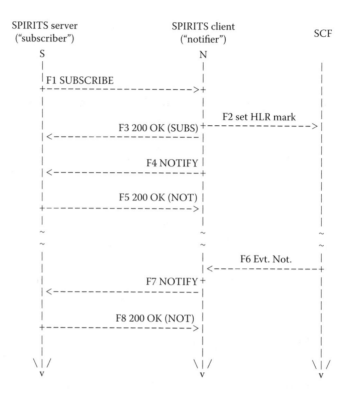

Figure 4   Sample call flow

In F1 of Figure 4, the subscriber indicates an interest in receiving a notification when a mobile user registers with the cellular network. The cellular network notes this event (F2) and confirms the subscription (F3-F5). When the mobile user turns on her cell phone and registers with the network, this event is detected (F6). The cellular network then sends out a notification to the subscriber informing it of this event (F7-F8).

We present the details of the call flow next.

In F1, the subscriber subscribes to the registration event (REG) of a cellular phone number.

F1: S->N

SUBSCRIBE sip:myprovider.com SIP/2.0
From: <sip:vkg@example.com>;tag=8177-afd-991
To: <sip:16302240216@myprovider.com>

```
CSeq: 18992 SUBSCRIBE
Call-ID: 3329as77@host.example.com
Contact: <sip:vkg@host.example.com>
Via: SIP/2.0/UDP host.example.com;branch=z9hG4bK776asdhdsa8
Expires: 3600
Event: spirits-user-prof
Allow-Events: spirits-INDPs, spirits-user-prof
Accept: application/spirits-event+xml
Content-Type: application/spirits-event+xml
Content-Length: ...
```

```
<?xml version="1.0" encoding="UTF-8"?>
<spirits-event xmlns="urn:ietf:params:xml:ns:spirits-1.0">
   <Event type="userprof" name="REG">
     <CalledPartyNumber>6302240216</CalledPartyNumber>
   </Event>
</spirits-event>
```

The subscription reaches the notifier which authenticates the request (not shown) and interacts with the SCF to update the subscriber database for this event. The notifier sends out a successful response to the subscription:

F3: N->S

```
SIP/2.0 200 OK
From: <sip:vkg@example.com>;tag=8177-afd-991
To: <sip:16302240216@myprovider.com>;tag=SPIRITS-REG-
16302240216
CSeq: 18992 SUBSCRIBE
Call-ID: 3329as77@host.example.com
Contact: <sip:notifier.myprovider.com>
Via: SIP/2.0/UDP host.example.com;branch=z9hG4bK776asdhdsa8
Expires: 3600
Allow-Events: spirits-INDPs, spirits-user-prof
Accept: application/spirits-event+xml
Content-Length: 0
```

The notifier also sends out a NOTIFY request confirming the subscription:

F4: N->S

NOTIFY sip:vkg@host.example.com SIP/2.0
To: <sip:vkg@example.com>;tag=8177-afd-991
From: <sip:16302240216@myprovider.com>;tag=SPIRITS-REG-16302240216
CSeq: 9121 NOTIFY
Call-ID: 3329as77@host.example.com
Contact: <sip:notifier.myprovider.com>
Subscription-State: active
Via: SIP/2.0/UDP notifier.myprovider.com;branch=z9hG4bK7007-091a
Allow-Events: spirits-INDPs, spirits-user-prof
Accept: application/spirits-event+xml
Content-Length: 0

The subscriber confirms the receipt of the NOTIFY request:

F5: S->N

SIP/2.0 200 OK
To: <sip:vkg@example.com>;tag=8177-afd-991
From: <sip:16302240216@myprovider.com>;tag=SPIRITS-REG-16302240216
CSeq: 9121 NOTIFY
Call-ID: 3329as77@host.example.com
Contact: <sip:vkg@host.example.com>
Via: SIP/2.0/UDP notifier.myprovider.com;branch=z9hG4bK7007-091a
Content-Length: 0

In F6, the mobile user identified by the PSTN number "6302240216" turns the mobile phone on, thus causing it to register with the cellular network. The cellular network detects this event, and since a subscriber has indicated an interest in receiving a notification of this event, a SIP NOTIFY request is transmitted towards the subscriber:

F7: N->S

NOTIFY sip:vkg@host.example.com SIP/2.0
To: <sip:vkg@example.com>;tag=8177-afd-991

From: <sip:16302240216@myprovider.com>;tag=SPIRITS-REG-
16302240216
CSeq: 9122 NOTIFY
Call-ID: 3329as77@host.example.com
Contact: <sip:notifier.myprovider.com>
Subscription-State: terminated;reason=fired
Via: SIP/2.0/UDP notifier.myprovider.com;branch=z9hG4bK7yi-p12
Event: spirits-user-prof
Allow-Events: spirits-INDPs, spirits-user-prof
Accept: application/spirits-event+xml
Content-Type: application/spirits-event+xml
Content-Length: ...

```
<?xml version="1.0" encoding="UTF-8"?>
<spirits-event xmlns="urn:ietf:params:xml:ns:spirits-1.0">
   <Event type="userprof" name="REG">
      <CalledPartyNumber>6302240216</CalledPartyNumber>
      <Cell-ID>45987</Cell-ID>
   </Event>
</spirits-event>
```

The subscriber receives the notification and acknowledges it by sending a response:

F8: S->N

SIP/2.0 200 OK
To: <sip:vkg@example.com>;tag=8177-afd-991
From: <sip:16302240216@myprovider.com>;tag=SPIRITS-REG-
16302240216
CSeq: 9122 NOTIFY
Call-ID: 3329as77@host.example.com
Via: SIP/2.0/UDP notifier.myprovider.com;branch=z9hG4bK7yi-p12
Content-Length: 0

Note that once the subscriber has received this notification, it can execute appropriate services. In this particular instance, an appropriate service may consist of the subscriber acting as a composer of a presence service and turning the presence status of the user associated with the phone number "6302240216" to "on". Also note in F7 that the notifier included a Cell ID in the notification.

The Cell ID can be used as a basis for location specific services; however, a discussion of such services is out of the scope of this document.

6.15. Use of URIs to retrieve state

The "spirits-user-prof" package MUST NOT use URIs to retrieve state. It is expected that most state information for this package is compact enough to fit in a SIP message. However, to err on the side of caution, implementations MUST follow the convention outlined in Section 18.1.1 of [5] and use a congestion controlled transport if the size of the request is within 200 bytes of the path MTU if known, or if the request size is larger than 1300 bytes and the path MTU is unknown.

7. IANA Considerations

This document calls for IANA to:

- register two new SIP Event Packages per [3].
- register a new MIME type per [20].
- register a new namespace URN per [16].
- register a new XML schema per [16].

7.1. Registering event packages

Package Name: spirits-INDPs
Type: package
Contact: Vijay K. Gurbani, vkg@lucent.com
Reference: RFC 3910
Package Name: spirits-user-prof
Type: package
Contact: Vijay K. Gurbani, vkg@lucent.com
Reference: RFC 3910

7.2. Registering MIME type

MIME media type name: application
MIME subtype name: spirits-event+xml
Mandatory parameters: none
Optional parameters: charset (same semantics of charset parameter in application/xml [9])
Encoding considerations: same as considerations outlined for application/xml in [9].
Security considerations: Section 10 of [9] and Section 8 of this document.

Interoperability considerations: none.

Published specifications: this document.

Applications which use this media type: SPIRITS aware entities which adhere to this document.

Additional information:

　　Magic number(s): none.

　　File extension(s): none.

　　Macintosh file type code(s): none.

　　Object Identifier(s) or OID(s): none.

Person and email address for further information: Vijay K. Gurbani, <vkg@lucent.com>

Intended usage: Common

Author/Change controller: The IETF

7.3. *Registering URN*

URI

　　urn:ietf:params:xml:ns:spirits-1.0

Description

　　This is the XML namespace URI for XML elements defined by this document. Such elements describe the SPIRITS information in the "application/ spirits-event+xml" content type.

Registrant Contact

　　IESG.

XML

```
BEGIN
  <?xml version="1.0"?>
  <!DOCTYPE html PUBLIC "-//W3C//DTD XHTML Basic 1.0//EN"
        "http://www.w3.org/TR/xhtml-basic/xhtml-basic10.dtd">
  <html xmlns="http://www.w3.org/1999/xhtml">
  <head>
   <meta http-equiv="content-type"
     content="text/html;charset=utf-8"/>
   <title>Namespace for SPIRITS-related information</title>
  </head>
  <body>
   <h1>Namespace for SPIRITS-related information</h1>
   <h2>application/spirits-event+xml</h2>
   <p>See <a href="[[[URL of published RFC]]]">RFC3910</a>.</p>
  </body>
  </html>
END
```

7.4. Registering XML schema

URI
  urn:ietf:params:xml:schema:spirits-1.0
Description
  XML base schema for SPIRITS entities.
Registrant Contact
  IESG.
XML
  Please see XML schema definition in Section 9 of this document.

8. Security Considerations

This section focuses on security considerations which are unique to SPIRITS. SIP security mechanisms are discussed in detail in the core SIP specification [5] and are outside the scope of this document. SPIRITS security mechanisms are based on and strengthen SIP security [5], for example, SPIRITS mandates the support of S/MIME. Beyond that, any other security solutions specified in [5], i.e., TLS or HTTP Digest authentication, may be utilized by SPIRITS operators.

As outlined in Chapter 9 (Security Consideration) of RFC3298 [4], the following security aspects are applicable to the SPIRITS protocol:

- Authentication
- Integrity
- Confidentiality
- Non-repudiation

The SPIRITS architecture in Figure 1 contains 5 interfaces — A, B, C, D, and E. Of these, only two — B and C — are of interest to SPIRITS. Interfaces A and E are PINT interfaces and are thus assumed secured by the PINT RFC [8]. Security for interface D is out of scope in this document since it deals primarily with the PSTN infrastructure. We will discuss security aspects on interfaces B and C predicated on certain assumptions.

A driving assumption for SPIRITS security is that the SPIRITS gateway is owned by the same PSTN operator that owns the SPIRITS notifier. Thus, it is attractive to simply relegate security of interface C to the PSTN operator, and in fact, there are merits to doing just that since interface C crosses the IP Network and PSTN boundaries. However, a closer inspection reveals that both interfaces B and C transmit the SPIRITS protocol; thus, any security arrangement we arrive at for interface B can be suitably applied to interface C as well. This makes it possible to secure interface

C in case the SPIRITS gateway is not owned by the same PSTN operator that owns the SPIRITS notifier.

The ensuing security discussion assumes that the SPIRITS subscriber is communicating directly to the SPIRITS notifier (and vice-versa) and specifies a security apparatus for this arrangement. However, the same apparatus can be used to secure the communication between a SPIRITS subscriber and an intermediary (like the SPIRITS gateway), and the same intermediary and the SPIRITS notifier.

Confidentiality of the SPIRITS protocol is essential since the information carried in the protocol data units is of a sensitive nature and may lead to privacy concerns if revealed to non-authorized parties. The communication path between the SPIRITS notifier and the SPIRITS subscriber should be secured through S/MIME [18] to alleviate privacy concerns, as is described in the Security Consideration section of the core SIP specification [5].

> S/MIME is an end-to-end security mechanism which encrypts the SPIRITS bodies for transit across an open network. Intermediaries need not be cognizant of S/MIME in order to route the messages (routing headers travel in the clear).

S/MIME provides all the security aspects for SPIRITS outlined at the beginning of this section: authentication, message integrity, confidentiality, and non-repudiation. Authentication properties provided by S/MIME would allow the recipient of a SPIRITS message to ensure that the SPIRITS payload was generated by an authorized entity. Encryption would ensure that only those SPIRITS entities possessing a particular decryption key are capable of inspecting encapsulated SPIRITS bodies in a SIP request.

All SPIRITS endpoints MUST support S/MIME signatures (CMS Signed-Data) and MUST support encryption (CMS EnvelopedData).

If the B and C interfaces are owned by the same PSTN operator, it is possible that public keys will be installed in the SPIRITS endpoints.

S/MIME supports two methods — issuerAndSerialNumber and subject-KeyIdentifier — of naming the public key needed to validate a signature. Between these, subjectKeyIdentifier works with X.509 certificates and other schemes as well, while issuerAndSerialNumber works with X.509 certificates only. If the administrator configures the necessary public keys, providing integrity through procedural means, then S/MIME can be used without X.509 certificates.

All requests (and responses) between SPIRITS entities MUST be encrypted.

When a request arrives at a SPIRITS notifier from a SPIRITS subscriber, the SPIRITS notifier MUST authenticate the request. The subscription (or registration) from a SPIRITS subscriber MUST be rejected if the authentication fails. If the SPIRITS subscriber successfully authenticated itself to the SPIRITS notifier, the SPIRITS notifier MUST, at the very least, ensure that the SPIRITS subscriber is indeed allowed to receive notifications of the events it is subscribing to.

> Note that this document does not proscribe how the SPIRITS-notifier achieves this. In practice, it could be through access control lists (ACL) that are populated by a service management system in the PSTN, or through a web interface of some sort.

Requests from the SPIRITS notifier to the SPIRITS subscribers MUST also be authenticated, lest a malicious party attempts to fraudulently pose as a SPIRITS notifier to hijack a session.

9. XML schema definition

The SPIRITS payload is specified in XML; this section defines the base XML schema for documents that make up the SPIRITS payload. All SPIRITS entities that transport a payload characterized by the MIME type "application/spirits-event+xml" MUST support documents corresponding to the base schema below.

Multiple versions of the base schema are not expected; rather, any additional functionality (e.g., conveying new PSTN events) must be accomplished through the definition of a new XML namespace and a corresponding schema. Elements from the new XML namespace will then co-exist with elements from the base schema in a document.

```
<xs:schema targetNamespace="urn:ietf:params:xml:ns:spirits-1.0"
    xmlns:tns="urn:ietf:params:xml:ns:spirits-1.0"
    xmlns:xs="http://www.w3.org/2001/XMLSchema"
    elementFormDefault="qualified"
    attributeFormDefault="unqualified">

 <!-- This import brings in the XML language attribute xml:lang-->
 <xs:import namespace="http://www.w3.org/XML/1998/namespace"
        schemaLocation="http://www.w3.org/2001/xml.xsd"/>

 <xs:annotation>
    <xs:documentation xml:lang="en">
       Describes SPIRITS events.
    </xs:documentation>
 </xs:annotation>

 <xs:element name="spirits-event" type="tns:SpiritsEventType"/>
```

```
<xs:complexType name="SpiritsEventType">
  <xs:sequence>
   <xs:element name="Event" type="tns:EventType" minOccurs="1"
      maxOccurs="unbounded"/>
    <xs:any namespace="##other" processContents="lax"
      maxOccurs="unbounded"/>
  </xs:sequence>
</xs:complexType>

<xs:complexType name="EventType">
  <xs:sequence>
    <xs:element name="CalledPartyNumber" type="xs:token"
      minOccurs="0" maxOccurs="1"/>
    <xs:element name="CallingPartyNumber" type="xs:token"
      minOccurs="0" maxOccurs="1"/>
    <xs:element name="DialledDigits" type="xs:token"
      minOccurs="0" maxOccurs="1"/>
    <xs:element name="Cell-ID" type="xs:token"
      minOccurs="0" maxOccurs="1"/>
    <xs:element name="Cause" type="tns:CauseType"
      minOccurs="0" maxOccurs="1"/>
  </xs:sequence>
  <xs:attribute name="type" type="tns:PayloadType"
    use="required"/>
  <xs:attribute name="name" type="tns:EventNameType"
    use="required"/>
  <xs:attribute name="mode" type="tns:ModeType"
    use="optional" default="N"/>
</xs:complexType>
<xs:simpleType name="PayloadType">
   <!-- The <spirits-event> will contain either a list of -->
   <!-- INDPs events or a list of userprof events -->
   <xs:restriction base="xs:string">
    <xs:enumeration value="INDPs"/>
    <xs:enumeration value="userprof"/>
   </xs:restriction>
</xs:simpleType>

<xs:simpleType name="EventNameType">
   <xs:restriction base="xs:string">
    <!-- These are the call related events (DPs). If the -->
    <!-- PaylaodType is "INDPs", then the value of the "name"-->
    <!-- attribute is one of these; example -->
```

```
    <!-- <spirits-event type="INDPs" name="OCI"> -->
    <xs:enumeration value="OAA"/>
    <xs:enumeration value="OCI"/>
    <xs:enumeration value="OAI"/>
    <xs:enumeration value="OA"/>
    <xs:enumeration value="OTS"/>
    <xs:enumeration value="ONA"/>
    <xs:enumeration value="OCPB"/>
    <xs:enumeration value="ORSF"/>
    <xs:enumeration value="OMC"/>
    <xs:enumeration value="OAB"/>
    <xs:enumeration value="OD"/>
    <xs:enumeration value="TA"/>
    <xs:enumeration value="TMC"/>
    <xs:enumeration value="TAB"/>
    <xs:enumeration value="TD"/>
    <xs:enumeration value="TAA"/>
    <xs:enumeration value="TFSA"/>
    <xs:enumeration value="TB"/>
    <!-- These are the non-call related events. If the -->
    <!-- PayloadType is "user-prof", then the value of the -->
    <!-- "name" attribute is one of these; example -->
    <!-- <spirits-event type="userprof" name="LUDV"> -->
    <xs:enumeration value="LUSV"/>
    <xs:enumeration value="LUDV"/>
    <xs:enumeration value="REG"/>
    <xs:enumeration value="UNREGMS"/>
    <xs:enumeration value="UNREGNTWK"/>
   </xs:restriction>
 </xs:simpleType>

 <xs:simpleType name="ModeType">
   <!-- One of two values: "N"otification or "R"equest -->
   <xs:restriction base="xs:string">
    <xs:enumeration value="N"/>
    <xs:enumeration value="R"/>
   </xs:restriction>
 </xs:simpleType>

 <xs:simpleType name="CauseType">
   <xs:restriction base="xs:string">
    <xs:enumeration value="Busy"/>
    <xs:enumeration value="Unreachable"/>
```

```
      </xs:restriction>
    </xs:simpleType>
  </xs:schema>
```

10. Acknowledgements

The authors are grateful to participants in the SPIRITS WG for the discussion that contributed to this work. These include J-L. Bakker, J. Bjorkner, J. Buller, J-E. Chapron, B. Chatras, O. Cleuziou, L. Conroy, R. Forbes, F. Haerens, J. Humphrey, J. Kozik, W. Montgomery, S. Nyckelgard, M. O'Doherty, A. Roach, J. Rosenberg, H. Sinnreich, L. Slutsman, D. Varney, and W. Zeuch. The authors also acknowledge Steve Bellovin, Allison Mankin and Jon Peterson for help provided on the Security section.

11. Acronyms

ACL	Access Control List
CS	Capability Set
DP	Detection Point
DTD	Document Type Definition
EDP	Event Detection Point
EDP-N	Event Detection Point "Notification"
EDP-R	Event Detection Point "Request"
IANA	Internet Assigned Numbers Authority
ICW	Internet Call Waiting
IMSI	International Mobile Subscriber Identity
IN	Intelligent Network
INAP	Intelligent Network Application Protocol
IP	Internet Protocol
ISP	Internet Service Provider
ITU	International Telecommunications Union
MIME	Multipurpose Internet Mail Extensions
MS	Mobile Station (or Mobile Subscriber)
OBCSM	Originating Basic Call State Model
PIC	Point In Call
PINT	PSTN/Internet Interworking
PSTN	Public Switched Telephone Network
SCF	Service Control Function
SCP	Service Control Point
SDP	Session Description Protocol
SIP	Session Initiation Protocol

SIP-T	SIP for Telephones
SPIRITS	Services in the PSTN/IN Requesting InTernet Services
SSF	Service Switching Function
SSP	Service Switching Point
STD	State Transition Diagram
TBCSM	Terminating Basic Call State Model
TDP	Trigger Detection Point
TDP-N	Trigger Detection Point "Notification"
TDP-R	Trigger Detection Point "Request"
TLS	Transport Layer Security
UA	User Agent
VLR	Visitor Location Register
WIN	Wireless Intelligent Network
XML	Extensible Markup Language

12. References

12.1. *Normative References*

[1] Slutsman, L., Faynberg, I., Lu, H., and M. Weissman, "The SPIRITS Architecture", RFC 3136, June 2001.

[2] Bradner, S., "Key words for use in RFCs to Indicate Requirement Levels", BCP 14, RFC 2119, March 1997.

[3] Roach, A., "Session Initiation Protocol (SIP)-Specific Event Notification", RFC 3265, June 2002.

[4] Faynberg, I., Gato, J., Lu, H., and L. Slutsman, "Service in the Public Switched Telephone Network/Intelligent Network (PSTN/IN) Requesting InTernet Service (SPIRITS) Protocol Requirements", RFC 3298, August 2002.

[5] Rosenberg, J., Schulzrinne, H., Camarillo, G., Johnston, A., Peterson, J., Sparks, R., Handley, M., and E. Schooler, "SIP: Session Initiation Protocol", RFC 3261, June 2002.

12.2. *Informative References*

[6] M. Unmehopa, K. Vemuri, A. Brusilovsky, E. Dacloush, A. Zaki, F. Haerens, J-L. Bakker, B. Chatras, and J. Dobrowolski, "On selection of IN parameters to be carried by the SPIRITS Protocol", Work In Progress, January 2003.

[7] Intelligent Network Capability Set 2. ITU-T, Recommendation Q.1228.

[8] Petrack, S. and L. Conroy, "The PINT Service Protocol: Extensions to SIP and SDP for IP Access to Telephone Call Services", RFC 2848, June 2000.

[9] Murata, M., St.Laurent, S., and D. Kohn, "XML Media Types", RFC 3023, January 2001.

[10] Lu, H., Faynberg, I., Voelker, J., Weissman, M., Zhang, W., Rhim, S., Hwang, J., Ago, S., Moeenuddin, S., Hadvani, S., Nyckelgard, S., Yoakum, J., and L. Robart, "Pre-Spirits Implementations of PSTN-initiated Services", RFC 2995, November 2000.

[11] Rosenberg, J. and H. Schulzrinne, "Session Initiation Protocol (SIP): Locating SIP Servers", RFC 3263, June 2002.

[12] Thompson, H., Beech, D., Maloney, M. and N. Mendelsohn, "XML Schema Part 1: Structures", W3C REC REC-xmlschema-1-20010502, May 2001. <http://www.w3c.org/XML/>.

[13] "Interface recommendations for intelligent network capability set 3: SCF-SSF interface", ITU-T Recommendation Q.1238.2, June 2000.

[14] Moats, R., "URN Syntax", RFC 2141, May 1997.

[15] Moats, R., "A URN Namespace for IETF Documents", RFC 2648, August 1999.

[16] Mealling, M., "The IETF XML Registry", BCP 81, RFC 3688, January 2004.

[17] Tim Bray, Dave Hollander, and Andrew Layman, "Namespaces in XML", W3C recommendation: xml-names, 14th January 1999, <http://www.w3.org/ TR/REC-xml-names>.

[18] Ramsdell, B., "Secure/Multipurpose Internet Mail Extensions (S/MIME) Version 3.1 Message Specification", RFC 3851, July 2004.

[19] Faynberg, I., L. Gabuzda, M. Kaplan, and N.Shah, "The Intelligent Network Standards: Their Application to Services", McGraw-Hill, 1997.

[20] Freed, N. and N. Borenstein, "Multipurpose Internet Mail Extensions (MIME) Part One: Format of Internet Message Bodies", RFC 2045, November 1996.

Freed, N. and N. Borenstein, "Multipurpose Internet Mail Extensions (MIME) Part Two: Media Types", RFC 2046, November 1996.

Moore, K., "MIME (Multipurpose Internet Mail Extensions) Part Three: Message Header Extensions for Non-ASCII Text", RFC 2047, November 1996.

Freed, N., Klensin, J., and J. Postel, "Multipurpose Internet Mail Extensions (MIME) Part Four: Registration Procedures", BCP 13, RFC 2048, November 1996.

Freed, N. and N. Borenstein, "Multipurpose Internet Mail Extensions (MIME) Part Five: Conformance Criteria and Examples", RFC 2049, November 1996.

13. Contributors

Kumar Vemuri
Lucent Technologies, Inc.
2000 Naperville Rd.
Naperville, IL 60566
USA
EMail: vvkumar@lucent.com

14. Authors' Addresses

Vijay K. Gurbani, Editor
2000 Lucent Lane
Rm 6G-440
Naperville, IL 60566
USA
EMail: vkg@lucent.com

Alec Brusilovsky
2601 Lucent Lane
Lisle, IL 60532-3640
USA
EMail: abrusilovsky@lucent.com

Igor Faynberg
Lucent Technologies, Inc.
101 Crawfords Corner Rd.
Holmdel, NJ 07733
USA
EMail: faynberg@lucent.com

Jorge Gato
Vodafone Espana
Isabel Colbrand, 22
28050 Madrid
Spain
EMail: jorge.gato@vodafone.com

Hui-Lan Lu
Bell Labs/Lucent Technologies
Room 4C-607A, 101 Crawfords Corner Road
Holmdel, New Jersey, 07733
Phone: (732) 949-0321
EMail: huilanlu@lucent.com

Musa Unmehopa
Lucent Technologies, Inc.
Larenseweg 50,
Postbus 1168
1200 BD, Hilversum,
The Netherlands
EMail: unmehopa@lucent.com

15. Full Copyright Statement

Intellectual Property

Acknowledgement

Funding for the RFC Editor function is currently provided by the Internet Society.

Appendix B

Network Working Group
Request for Comments: 3976
Category: Informational

V. K. Gurbani
Lucent Technologies, Inc.
F. Haerens
Alcatel Bell
V. Rastagi
Wipro Technologies
January 2005

Interworking SIP and Intelligent Network (IN) Applications

Status of this Memo

Copyright Notice

IESG Note

This RFC is not a candidate for any level of Internet Standard. The IETF disclaims any knowledge of the fitness of this RFC for any purpose, and in particular notes that the decision to publish is not based on IETF review for such things as security, congestion control, or inappropriate interaction with deployed protocols. The RFC Editor has chosen to publish this document at its discretion. Readers of this document should exercise caution in evaluating its value for implementation and deployment. See RFC 3932 for more information.

Abstract

Public Switched Telephone Network (PSTN) services such as 800-number routing (freephone), time-and-day routing, credit-card calling, and virtual private network (mapping a private network number into a public number) are realized by the Intelligent Network (IN). This document addresses means to support existing IN services from Session Initiation Protocol (SIP) endpoints for an IP-host-to-phone call. The call request is originated on a SIP endpoint, but the services to the call are provided by the data and procedures resident in the PSTN/IN. To provide IN services in a transparent manner to SIP endpoints, this document describes the mechanism for interworking SIP and Intelligent Network Application Part (INAP).

TABLE OF CONTENTS

1.   Introduction

PSTN services such as 800-number routing (freephone), time-and-day routing, credit-card calling, and virtual private network (mapping a private network number into a public number) are realized by the Intelligent Network. IN is an architectural concept for the real-time execution of network services and customer applications [1]. IN is, by design, de-coupled from the call processing component of the PSTN. In this document, we describe the means to leverage this decoupling to provide IN services from SIP-based entities.

First, we will explain the basics of IN. Figure 1 shows a simplified IN architecture, in which telephone switches called Service Switching Points (SSPs) are connected via a packet network called Signaling System No. 7 (SS7) to Service Control Points (SCPs), which are general purpose computers. At certain points in a call, a switch can interrupt a call and request instructions from an SCP on how to proceed with the call. The points at which a call can be interrupted are standardized within the Basic Call State Model (BCSM) [1, 2]. The BCSM models contain two processes, one each for the originating and terminating part of a call.

When the SCP receives a request for instructions, it can reply with a single response, such as a simple number translation augmented by criteria like time of day or day of week, or, in turn, initiate a complex dialog with

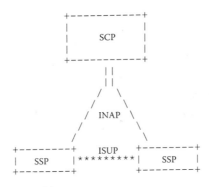

Figure 1.   Simplified IN Architecture

the switch. The situation is further complicated by the necessity to engage other specialized devices that collect digits, play recorded announcements, perform text-to-speech or speech-to-text conversions, etc. (These devices are not discussed here.) The related protocol, as well as the BCSM, is standardized by the ITU-T and known as the Intelligent Network Application Part protocol (INAP) [4]. Only the protocol, not an SCP API, has been standardized.

The overall objective is to ensure that IN control of Voice over IP (VoIP) services in networks can be readily specified and implemented by adapting standards and software used in the present networks. This approach leads to services that function the same when a user connects to present or future networks, simplifies service evolution from present to future, and leads to more rapid implementation.

The rest of this document is organized as follows: Section 2 contains the architectural model of an IN aware SIP entity. Section 3 provides some issues to be taken into account when performing SIP/IN interworking (SIN). Section 4 discusses the IN service control based on the SIN approach. The technique outlined in this document focuses on the call models of IN and the SIP protocol state machine; Section 5 thus establishes a complete mapping between the two state machines that allows access to IN services from SIP endpoints. Section 6 includes call flows of IN services executing on SIP endpoints. These services are readily enabled by the technique described in this document. Finally, Section 7 covers security aspects of SIN.

List of Acronyms

B2BUA	Back-to-Back User Agent
BCSM	Basic Call State Model
CCF	Call Control Function
DP	Detection Point
DTMF	Dual Tone Multi-Frequency
IN	Intelligent Network
INAP	Intelligent Network Application Part
IP	Internet Protocol
ITU-T	International Telecommunications Union – Telecommunications Standardization Sector
O_BCSM	Originating Basic Call State Model
PIC	Point in Call
PSTN	Public Switched Telephone Network
RTP	Real Time Protocol
R-URI	Request URI

SCF	Service Control Function
SCP	Service Control Point
SIGTRAN	Signal Transport Working Group in IETF
SIN	SIP/IN Interworking
SIP	Session Initiation Protocol
SS7	Signaling System No. 7
SSF	Service Switching Function
SSP	Service Switching Point
T_BCSM	Terminating Basic Call State Model
UA	User Agent
UAC	User Agent Client
UAS	User Agent Server
VoIP	Voice over IP
VPN	Virtual Private Network

2.   Access to IN-Services from a SIP Entity

The intent of this document is to provide the means to support existing IN-based applications in a SIP [3] environment. One way to gain access to IN services transparently from SIP (e.g., through the same detection points (DPs) and point-in-call (PIC) used by traditional switches) is to map the SIP protocol state machine to the IN call models [1].

From the viewpoint of IN elements such as the SCP, the request's origin from a SIP entity rather than a call processing function on a traditional switch is immaterial. Thus, it is important that the SIP entity be able to provide the same features as the traditional switch, including operating as an SSP for IN features. The SIP entity should also maintain call state and trigger queries to IN-based services, as do traditional switches.

This document does not intend to specify which SIP entity shall operate as an SSP; however, for the sake of completeness, it should be mentioned that this task should be performed by SIP entities at (or near) the core of the network rather than at the SIP end points themselves. To that extent, SIP entities such as proxy servers and Back-to-Back user agents (B2BUAs) may be employed. Generally speaking, proxy servers can be used for IN services that occur during a call setup and teardown. For IN services requiring specialized media handling (such as DTMF detection) or specialized call control (such as placing parties on hold) B2BUAs will be required.

The most expeditious manner for providing existing IN services in the IP domain is to use the deployed IN infrastructure as often as possible. In SIP, the logical point to tap into for accessing existing IN services is either the user agents or one of the proxies physically closest to the user agent (and presumably in the same administrative domain). However, SIP entities do not run an IN call model; to access IN services transparently, the trick then is to overlay the state machine of the SIP entity with an IN layer so that call acceptance and routing is performed by the native state machine and so that services are accessed through the IN layer by using an IN call model. Such an IN-enabled SIP entity, operating in synchrony with the events occurring at the SIP transaction level and interacting with the IN elements (SCP), is depicted in Figure 2:

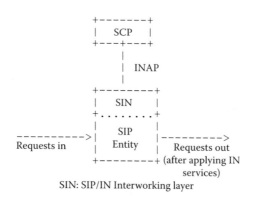

Figure 2.   SIP Entity Accessing IN Services

Section 5 proposes this mapping between the IN layer and the SIP protocol state machine. Essentially, a SIP entity exhibiting this mapping becomes a SIN-enabled SIP entity.

This document does not propose any extensions to SIP.

Figure 3 expands the SIP entity depicted in Figure 2 and further details the architecture model involving IN and SIP interworking. Events occurring at the SIP layer will be passed to the IN layer for service application. More specifically, since IN services deal with E.164 numbers, it is reasonable to assume that a SIN-enabled SIP entity that seeks to provide services on such a number will consult the IN layer for further processing, thus acting as a SIP-based SSP. The IN layer will proceed through its BCSM states and, at appropriate points in the call, will send queries to the SCP for call disposition. Once the disposition of the call has been determined, the SIP layer is informed and processes the transaction accordingly.

Note that the single SIP entity as modeled in this figure can in fact represent several different physical instances in the network as, for example, when one SIP entity is in charge of the terminal or access network/domain, and another is in charge of the interface to the Switched Circuit Network (SCN).

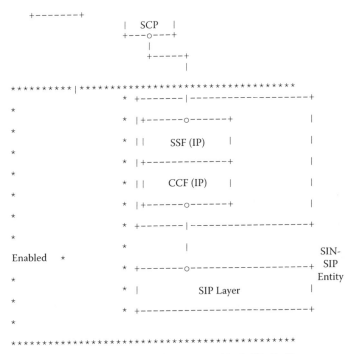

Figure 3.   Functional Architecture of a SIN-Enabled SIP Entity

The following architecture entities, used in Figure 3, are defined in the Intelligent Network standards:

> Service Switching Function (SSF): IN functional entity that interacts with call control functions.
> Call Control Function (CCF): IN functional entity that refers to call and connection handling in the classical sense (i.e., that of an exchange).

3.   Additional SIN Considerations

In working between Internet Telephony and IN-PSTN networks, the main issue is to translate between the states produced by the Internet Telephony signaling and those used in traditional IN environments. Such a translation entails attention to the considerations listed below.

3.1. The Concept of State in SIP

IN services occur within the context of a call, i.e., during call setup, call teardown, or in the middle of a call. SIP entities such as proxies, with which some of these services may be realized, typically run in transaction-stateful (or stateless) mode. In this mode, a SIP proxy that proxied the initial INVITE is not guaranteed to receive a subsequent request, such as a BYE. Fortunately, SIP has primitives to force proxies to run in a call-stateful mode; namely, the Record-Route header. This header forces the user agent client (UAC) and user agent server (UAS) to create a "route set" that consists of all intervening proxies through which subsequent requests must traverse. Thus SIP proxies must run in call-stateful mode in order to provide IN services on behalf of the UAs.

A B2BUA is another SIP element in which IN services can be realized. As a B2BUA is a true SIP UA, it maintains complete call state and is thus capable of providing IN services.

3.2. Relationship between SCP and a SIN-Enabled SIP Entity

In the architecture model proposed in this document, each SIN-enabled SIP entity is pre-configured to communicate with one logical SCP server, using whatever communication mechanism is appropriate. Different SIP servers (e.g., those in different administrative domains) may communicate with different SCP servers, so that there is no single SCP server responsible for all SIP servers.

As Figures 1 and 2 depict, the IN-portion of the SIN-enabled SIP entity will communicate with the SCP. This interface between the IN call handling layer and the SCP is not specified by this document and, indeed, can be any one of the following, depending on the interfaces supported by the SCP: INAP over IP, INAP over SIGTRAN, or INAP over SS7.

This document is only applicable when SIP-controlled Internet telephony devices seek to operate with PSTN devices. The SIP UAs using this interface would typically appear together with a media gateway. This document is *not* applicable in an all-IP network and is not needed in cases where PSTN media gateways (not speaking SIP) need to communicate with SCPs.

3.3. SIP REGISTER and IN Services

SIP REGISTER provisions a SIP Proxy or SIP Registration server. The process is similar to the provisioning of an SCP/HLR in the switched circuit network. SCPs that provide VoIP based services can leverage this information directly.

However, this document neither endorses nor prohibits such an architecture and, in fact, considers it an implementation decision.

3.4. Support of Announcements and Mid-Call Signaling

Services in the IN such as credit-card calling typically play announcements and collect digits from the caller before a call is set up. Playing announcements and collecting digits require the manipulation of media streams. In SIP, proxies do not have access to the media data path. Thus, such services should be executed in a B2BUA.

Although the SIP specification [3] allows for end points to be put on hold during a call or for a change of media streams to take place, it does not have any primitives to transport other than mid-call control information. This may include transporting DTMF digits, for example. Extensions to SIP, such as the INFO method [5] or the SIP event notification extension [6], can be considered for services requiring mid-call signaling. Alternatively, DTMF can be transported in RTP itself [7].

4. The SIN Architecture

4.1. Definitions

The SIP architecture has the following functional elements defined in [3]:

- User agent client (UAC): The SIP functional entity that initiates a request.
- User agent server (UAS): The SIP functional entity that terminates a request by sending 0 or more provisional SIP responses and one final SIP response.
- Proxy server: An intermediary SIP entity that can act as both a UAS and a UAC. Acting as a UAS, it accepts requests from UACs, rewrites the Request-URI (R-URI), and, acting as a UAC, proxies the request to a downstream UAS. Proxies may retain significant call control state by inserting themselves in future SIP transactions beyond the initial INVITE.
- Redirect server: An intermediary SIP entity that redirects callers to alternate locations, after possibly consulting a location server to determine the exact location of the callee (as specified in the R-URI).
- Registrar: A SIP entity that accepts SIP REGISTER requests and maintains a binding from a high-level URL to the exact location for a user. This information is saved in some data-store that is also

accessible to a SIP Proxy and a SIP Redirect server. A Registrar is usually co-located with a SIP Proxy or a SIP Redirect server.

■ Outbound proxy: A SIP proxy located near the originator of requests. It receives all outgoing requests from a particular UAC, including those requests whose R-URIs identify a host other than the outbound proxy. The outbound proxy sends these requests, after any local processing, to the address indicated in the R-URI.

■ Back-to-Back UA (B2BUA): A SIP entity that receives a request and processes it as a UAS. It also acts as a UAC and generates requests to determine how the incoming request is to be answered. A B2BUA maintains complete dialog state and must participate in all requests sent within the dialog.

4.2. IN Service Control Based on the SIN Approach

Figure 4 depicts the possibility of IN service control based on the SIN approach. On both the originating and terminating ends, a SIN-capable SIP entity is assumed (it can be a proxy or a B2BUA). The "O SIP" entity is required for outgoing calls that require support for existing IN services. Likewise, on the callee's side (or terminating side), an equally configured entity ("T SIP") will be required to provide terminating side services. Note that the "O SIP" and "T SIP" entities correspond, respectively, to the IN O_BCSM and T_BCSM halves of the IN call model.

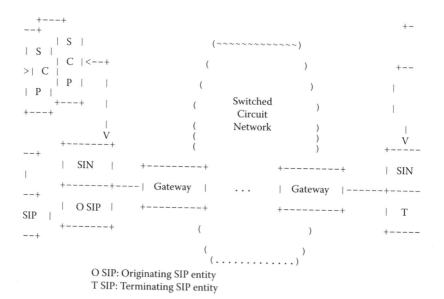

O SIP: Originating SIP entity
T SIP: Terminating SIP entity

Figure 4.   Overall SIN Architecture

5. Mapping of the SIP State Machine to the IN State Model

This section establishes the mapping of the SIP protocol state machine to the IN generic basic call state model (BCSM) [2], independent of any capability sets [8, 9]. The BCSM is divided into two halves: an originating call model (O_BCSM) and a terminating call model (T_BCSM). There are a total of 19 PICs and 35 DPs between both the halves (11 PICs and 21 DPs for O_BCSM; 8 PICs and 14 DPs for T_BCSM) [1]. The SSPs, SCPs, and other IN elements track a call's progress in terms of the basic call model. The basic call model provides a common context for communication about a call.

O_BCSM has 11 PICs:

O_NULL: Starting state; call does not exist yet.
AUTH_ORIG_ATTEMPT: Switch detects a call setup request.
COLLECT_INFO: Switch collects the dial string from the calling party.
ANALYZE_INFO: Complete dial string is translated into a routing address.
SELECT_ROUTE: Physical route is selected, based on the routing address.
AUTH_CALL_SETUP: Switch ensures the calling party is authorized to place the call.
CALL_SENT: Control of call sent to terminating side.
O_ALERTING: Switch waits for the called party to answer.
O_ACTIVE: Connection established; communications ensue.
O_DISCONNECT: Connection torn down.
O_EXCEPTION: Switch detects an exceptional condition.

T_BCSM has 8 PICS:

T_NULL: Starting state; call does not exist yet.
AUTH_TERM_ATT: Switch verifies whether the call can be sent to terminating party.
SELECT_FACILITY: Switch picks a terminating resource to send the call on.
PRESENT_CALL: Call is being presented to the called party.
T_ALERTING: Switch alerts the called party, e.g., by ringing the line.
T_ACTIVE: Connection established; communications ensue.
T_DISCONNECT: Connection torn down.
T_EXCEPTION: Switch detects an exceptional condition.

The state machine for O_BCSM and T_BCSM is provided in [1] on pages 98 and 103, respectively. This state machine will be used for subsequent discussion when the IN call states are mapped into SIP.

The next two sections contain the mapping of the SIP protocol state machine to the IN BCSMs. Explaining all PICs and DPs in an IN call model is beyond the scope of this document. It is assumed that the reader has some familiarity with the PICs and DPs of the IN call model. More information can be found in [1]. For a quick reference, Appendix A contains a mapping of the DPs to the SIP response codes as discussed in the next two sections.

5.1. *Mapping SIP Protocol State Machine to O_BCSM*

The 11 PICs of O_BCSM come into play when a call request (SIP INVITE message) arrives from an upstream SIP client to an originating SIN-enabled SIP entity running the IN call model. This entity will create an O_BCSM object and initialize it in the O_NULL PIC. The next seven IN PICs — O_NULL, AUTH_ORIG_ATT, COLLECT_INFO, ANALYZE_INFO, SELECT_ROUTE, AUTH_CALL_SETUP, and CALL_SENT — can all be mapped to the SIP "Calling" state.

Figure 5 provides a visual map from the SIP protocol state machine to the originating half of the IN call model. Note that control of the call shuttles between the SIP protocol machine and the IN O_BCSM call model while it is being serviced.

The SIP "Calling" protocol state has enough functionality to absorb the seven PICs as described below:

> O_NULL: This PIC is basically a fall through state to the next PIC, AUTHORIZE_ORIGINATION_ATTEMPT.
>
> AUTHORIZE_ORIGINATION_ATTEMPT: In this PIC, the IN layer has detected that someone wishes to make a call. Under some circumstances (e.g., if the user is not allowed to make calls during certain hours), such a call cannot be placed. SIP can authorize the calling party by using a set of policy directives configured by the SIP administrator. If the called party is authorized to place the call, the IN layer is instructed to enter the next PIC, COLLECT_INFO through DP 3 (Origination_Attempt_Authorized). If for some reason the call cannot be authorized, DP 2 (Origination_Denied) is processed, and control transfers to the SIP state machine. The SIP state machine must format and send a non-2xx final response (possibly 403) to the upstream entity.
>
> COLLECT_INFO: This PIC is responsible for collecting a dial string from the calling party and verifying the format of the string. If overlap dialing is being used, this PIC can invoke DP 4 (Collect_Timeout)

and transfer control to the SIP state machine, which will format and send a non-2xx final response (possibly a 484). If the dial string is valid, DP 5 (Collected_Info) is processed, and the IN layer is instructed to enter the next PIC, ANALYZE_INFO.

ANALYZE_INFO: This PIC is responsible for translating the dial string to a routing number. Many IN services, such as freephone, LNP (Local Number Portability), and OCS (Originating Call Screening) occur during this PIC. The IN layer can use the R-URI of the SIP INVITE request for analysis. If the analysis succeeds, the IN layer is instructed to enter the next PIC, SELECT_ROUTE.

If the analysis fails, DP 6 (Invalid_Info) is processed, and the control transfers to the SIP state machine, which will generate a non-2xx final response (possibly 400, 401, 403, 404, 405, 406, 410, 414, 415, 416, 485, or 488) and send it to the upstream entity.

SELECT_ROUTE: In the circuit-switched network, the actual physical route has to be selected at this point. The SIP analogue would be to determine the next hop SIP server. This could be chosen by a variety of means. For instance, if the Request URI in the incoming INVITE request is an E.164 number, the SIP entity can use a protocol like TRIP [10] to find the best gateway to egress the request onto the PSTN. If a successful route is selected, the IN call model moves to PIC AUTH_CALL_SETUP via DP 9 (Route_Selected). Otherwise, the control transfers to the SIP state machine via DP 8 (Route_Select_Failure), which will generate a non-2xx final response (possibly 488) and send it to the upstream entity.

AUTH_CALL_SETUP: Certain service features restrict the type of call that may originate on a given line or trunk. This PIC is the point at which relevant restrictions are examined. If no such restrictions are encountered, the IN call model moves to PIC CALL_SENT via DP 11 (Origination_Authorized). If a restriction is encountered that prohibits further processing of the call, DP 10 (Authorization_Failure) is processed, and control is transferred to the SIP state machine, which will generate a non-2xx final response (possibly 404, 488, or 502). Otherwise, DP 11 (Origination_Authorized) is processed, and the IN layer is instructed to enter the next PIC, CALL_SENT.

CALL_SENT: At this point, the request needs to be sent to the downstream entity. The IN layer waits for a signal confirming either that the call has been presented to the called party or that a called party cannot be reached for a particular reason. The control is transferred to the SIP state machine. The SIP state machine should now send the call to the next downstream server determined in

PIC SELECT_ROUTE. The IN call model now blocks until unblocked by the SIP state machine.

If the above seven PICs have been successfully negotiated, the SIN-enabled SIP entity now sends the SIP INVITE message to the next hop server. Further processing now depends on the provisional responses (if any) and the final response received by the SIP protocol state machine. The core SIP specification does not guarantee the delivery of 1xx responses; thus special processing is needed at the IN layer to transition to the next PIC (O_ALERTING) from the CALL_SENT PIC. The special processing needed for responses while the SIP state machine is in the "Proceeding" state and the IN layer is in the "CALL_SENT" state is described next.

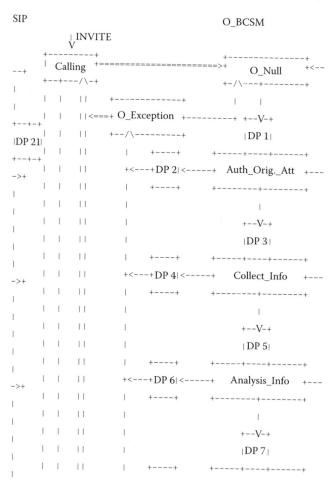

Figure 5. Mapping from SIP to O_BCSM

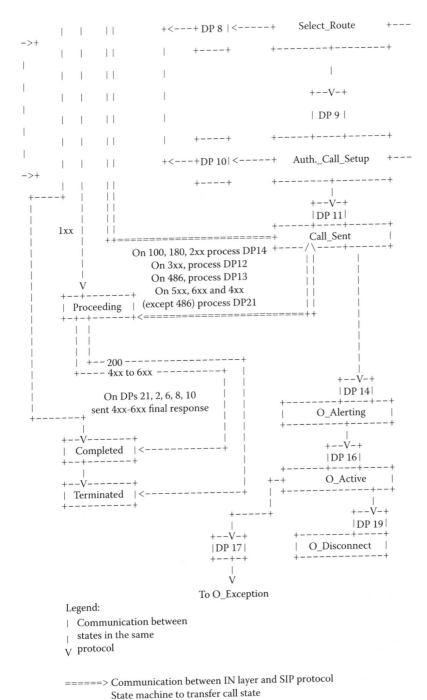

Figure 5. Mapping from SIP to O_BCSM (continued)

A 100 response received at the SIP state machine elicits no special behavior in the IN layer.

A 180 response received at the SIP entity enables the processing of DP 14 (O_Term_Seized), however, a state transition to O_ALERTING is not undertaken yet. Instead, the IN layer is instructed to remain in the CALL_SENT PIC until a final response is received.

A 2xx response received at the SIP entity enables the processing of DP 14 (O_Term_Seized), and the immediate transition to the next state, O_ALERTING (processing in O_ALERTING is described later).

A 3xx response received at the SIP entity enables the processing of DP 12 (Route_Failure). The IN call model from this point goes back to the SELECT_ROUTE PIC to select a new route for the contacts in the 3xx final response (not shown in Figure 5 for brevity).

A 486 (Busy Here) response received at the SIP entity enables the processing of DP 13 (O_Called_Party_Busy) and resources for the call are released at the IN call model.

If the SIN-enabled SIP entity gets a 4xx (except 486), 5xx, or 6xx final response, DP 21 (O_Calling_Party_Disconnect & O_Abandon) is processed and control passes to the SIP state machine. Since a call was not successfully established, both the IN layer and the SIP state machine can release resources for the call.

O_ALERTING - This PIC will be entered as a result of receiving 200-class response. Since a 200-class response to an INVITE indicates acceptance, this PIC is mostly a fall through to the next PIC, O_ACTIVE via DP 16 (O_Answer).

O_ACTIVE - At this point, the call is active. Once in this state, the call may get disconnected only when one of the following three events occur: (1) the network connection fails, (2) the called party disconnects the call, or (3) the calling party disconnects the call. If event (1) occurs, DP 17 (O_Connection_Failure) is processed and call control is transferred to the SIP protocol state machine. Since the network failed, there is not much sense in attempting to send a BYE request; thus, both the SIP protocol state machine and the IN call layer should release all resources associated with the call and initialize themselves to the null state. Event (2) results in the processing of DP 19 (O_DISCONNECT) and a move to the last PIC, O_DISCONNECT. Event (3) occurs if the calling party deliberately terminated the call. In this case, DP 21 (O_Abandon & O_Calling_Party_Disconnect) will be processed, and control will be passed to the SIP protocol state machine. The SIP protocol state machine must send a BYE request and wait for a final response.

The IN layer releases all of its resources and initializes itself to the null state.

O_DISCONNECT: When the SIP entity receives a BYE request, the IN layer is instructed to move to the last PIC, O_DISCONNECT via DP 19. A final response for the BYE is generated and transmitted by the SIP entity, and the call resources are freed by both the SIP protocol state machine and the IN layer.

5.2.   Mapping SIP Protocol State Machine to T_BCSM

The T_BCSM object is created when a SIP INVITE message makes its way to the terminating SIN-enabled SIP entity. This entity creates the T_BCSM object and initializes it to the T_NULL PIC.

Figure 6 provides a visual map from the SIP protocol state machine to the terminating half of the IN call model:

Figure 6.   Mapping from SIP to T_BCSM

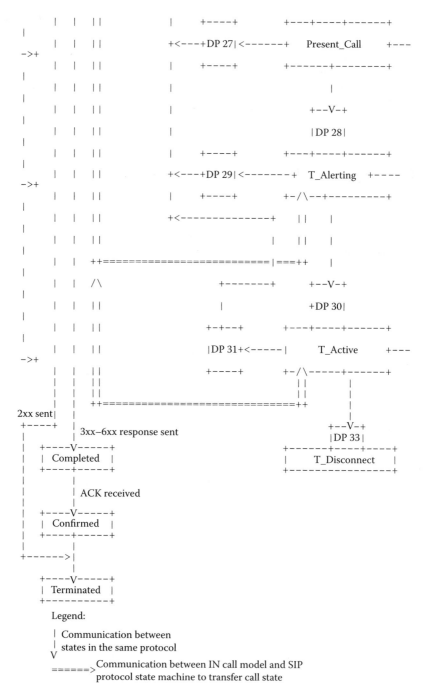

Legend:

| Communication between
| states in the same protocol
V

======> Communication between IN call model and SIP
protocol state machine to transfer call state

Figure 6.   Mapping from SIP to T_BCSM (continued)

The SIP "Proceeding" state has enough functionality to absorb the first five PICS — T_Null, Authorize_Termination_Attempt, Select_Facility, Present_Call, T_Alerting — as described below:

T_NULL: At this PIC, the terminating end creates the call at the IN layer. The incoming call results in the processing of DP 22, Termination_Attempt, and a transition to the next PIC, AUTHORIZE_TERMINATION_ATTEMPT, takes place.

AUTHORIZE_TERMINATION_ATTEMPT: At this PIC, it is ascertained that the called party wishes to receive the call and that the facilities of the called party are compatible with those of the calling party. If any of these conditions is not met, DP 23 (Termination_Denied) is invoked, and the call control is transferred to the SIP protocol state machine. The SIP protocol state machine can format and send a non-2xx final response (possibly 403, 405, 415, or 480). If the conditions of the PIC are met, processing of DP 24 (Termination_Authorized) is invoked, and a transition to the next PIC, SELECT_FACILITY, takes place.

SELECT_FACILITY: In circuit switched networks, this PIC is intended to select a line or trunk to reach the called party. As lines or trunks are not applicable in an IP network, a SIN-enabled SIP entity can use this PIC to interface with a PSTN gateway and select a line/trunk to route the call. If the called party is busy, or if a line/trunk cannot be seized, the processing of DP 25 (T_Called_Party_Busy) is invoked, and the call goes to the SIP protocol state machine. The SIP protocol state machine must format and send a non-2xx final response (possibly 486 or 600). If a line/trunk was successfully seized, the processing of DP 26 (Terminating_Resource_Available) is invoked, and a transition to the next PIC, PRESENT_CALL, takes place.

PRESENT_CALL: At this point, the call is being presented (via the ISUP ACM message, or Q.931 Alerting message, or simply by ringing a POTS phone). If there was an error presenting the call, the processing of DP 27 (Presentation_Failure) is invoked, and the call control is transferred to the SIP protocol state machine, which must format and send a non-2xx final response (possibly 480). If the call was successfully presented, the processing of DP 28 (T_Term_Seized) is invoked, and a transition to the next PIC, T_ALERTING, takes place.

T_ALERTING: At this point, the called party is being "alerted". Control now passes momentarily to the SIP protocol state machine so that it can generate and send a "180 Ringing" response to its peer. Furthermore, since network resources have been allocated for the call, timers are set to prevent indefinite holding of such resources. The expiration of the relevant timers results in the processing of DP 29 (T_No_Answer), and the call control is transferred to the SIP protocol state machine, which must format and send a non-2xx final response (possibly 408). If the called party answers, then DP 30 (T_Answer) is processed, followed by a transition to the next PIC, T_ACTIVE.

After the above five PICs have been negotiated, the rest are mapped as follows:

> T_ACTIVE: The call is now active. Once this state is reached, the call may become inactive under one of the following three conditions: (1) The network fails the connection, (2) the called party disconnects the call, or (3) the calling party disconnects the call. Event (1) results in the processing of DP 31 (T_Connection_Failure), and call control is transferred to the SIP protocol state machine. Since the network failed, there is little sense in attempting to send a BYE request; thus, both the SIP protocol state machine and the IN call layer should release all resources associated with the call and initialize themselves to the null state. Event (2) results in the processing of DP 33 (T_Disconnect) and a transition to the next PIC, T_DISCONNECT. Event (3) occurs at the receipt of a BYE request at the SIP protocol state machine (not shown in Figure 6). Resources for the call should be deallocated, and the SIP protocol state machine must send a 200 OK for the BYE request (not shown in Figure 6). T_DISCONNECT: In this PIC, the disconnect treatment associated with the called party's having disconnected the call is performed at the IN layer. The SIP protocol state machine sends out a BYE and awaits a final response for the BYE (not shown in Figure 6).

6. Examples of Call Flows

Two examples are provided here to show how SIP protocol state machine and the IN call model work synchronously with each other.

In the first example, a SIP UAC originates a call request destined to an 800 freephone number:

> INVITE sip:18005551212@example.com SIP/2.0
> From: sip:16305551212@example.net;tag=991-7as-66ff
> To: sip:18005551212@example.com

Via: SIP/2.0/UDP stn1.example.net
Call-ID: 67188121@example.net
CSeq: 1 INVITE

The request makes its way to the originating SIP network server running an IN call model. The SIP network server hands, at the very least, the To: field and the From: field to the IN layer for freephone number translation. The IN layer proceeds through its PICs and at the ANALYSE_INFO PIC consults the SCP for freephone translation. The translated number is returned to the SIP network server, which forwards the message to the next hop SIP proxy, with the freephone number replaced by the translated number:

INVITE sip:18475551212@example.com SIP/2.0
From: sip:16305551212@example.net;tag=991-7as-66ff
To: sip:18005551212@example.com
Via: SIP/2.0/UDP ext-stn2.example.net
Via: SIP/2.0/UDP stn1.example.net
Call-ID: 67188121@example.net
CSeq: 1 INVITE

In the next example, a SIP UAC originates a call request destined to a 900 number:

INVITE sip:19005551212@example.com SIP/2.0
From: sip:16305551212@example.net;tag=991-7as-66dd
To: sip:19005551212@example.com
Via: SIP/2.0/UDP stn1.example.net
Call-ID: 88112@example.net
CSeq: 1 INVITE

The request makes its way to the originating SIP network server running an IN call model. The SIP network server hands, at the very least, the To: field and the From: field to the IN layer for 900 number translation. The IN layer proceeds through its PICs and at the ANALYSE_INFO PIC consults the SCP for the translation. During the translation, the SCP detects that the originating party is not allowed to make 900 calls. It passes this information to the originating SIP network server, which informs the SIP UAC by using a SIP "403 Forbidden" response status code:

SIP/2.0 403 Forbidden
From: sip:16305551212@example.net;tag=991-7as-66dd
To: sip:19005551212@example.com;tag=78K-909II

Via: SIP/2.0/UDP stn1.example.net
Call-ID: 88112@example.net
CSeq: 1 INVITE

7. Security Considerations

Security considerations for SIN services cover both networks being used, namely, the PSTN and the Internet. SIN uses the security measures in place for both the networks. With reference to Figure 2, the INAP messages between the SCP and the SIN-enabled SIP entity must be secured by the signaling transport used between the SCP and the SIN-enabled entity. Likewise, the requests coming into the SIN-enabled SIP entity must first be authenticated and, if need be, encrypted as well, using the means and procedures defined in [3] for SIP requests.

8. References

8.1. Normative References

[1] I. Faynberg, L. Gabuzda, M. Kaplan, and N.Shah, "The Intelligent Network Standards: Their Application to Services," McGraw-Hill, 1997.

[2] ITU-T Q.1204 1993: Recommendation Q.1204, "Intelligent Network Distributed Functional Plane Architecture," International Telecommunications Union Standardization Section, Geneva.

[3] Rosenberg, J., Schulzrinne, H., Camarillo, G., Johnston, A., Peterson, J., Sparks, R., Handley, M., and E. Schooler, "SIP: Session Initiation Protocol", RFC 3261, June 2002.

8.2. Informative References

[4] ITU-T Q.1208: "General aspects of the Intelligent Network Application protocol"

[5] Donovan, S., "The SIP INFO Method", RFC 2976, October 2000.

[6] Roach, A.B., "Session Initiation Protocol (SIP)-Specific Event Notification", RFC 3265, June 2002.

[7] Schulzrinne, H. and S. Petrack, "RTP Payload for DTMF Digits, Telephony Tones and Telephony Signals", RFC 2833, May 2000.

[8] ITU-T Q.1218: "Interface Recommendation for Intelligent Network Capability Set 1".

[9] ITU-T Q.1228: "Interface Recommendation for Intelligent Network Capability Set 2".

[10] Rosenberg, J., Salama, H., and M. Squire, "Telephony Routing over IP (TRIP)", RFC 3219, January 2002.

APPENDIX A: Mapping of 4xx-6xx Responses in SIP to IN Detections Points

The mapping of error codes 4xx-6xx responses in SIP to the possible Detection Points in PIC Originating and Terminating Call Handling is indicated in the table below. The reason phrase in the 4xx-6xx response is reproduced from [3].

SIP response code	DP mapping to IN
200 OK	DP 14
3xx	DP 12
403 Forbidden	DP 2, DP 21
484 Address Incomplete	DP 4, DP 21
400 Bad Request	DP 6, DP 21
401 Unauthorized	DP 6, DP 21
403 Forbidden	DP 6, DP 21, DP 23
404 Not Found	DP 6, DP 21
405 Method Not Allowed	DP 6, DP 21, DP 23
406 Not Acceptable	DP 6, DP 21
408 Request Timeout	DP 29
410 Gone	DP 6, DP 21
414 Request-URI Too Long	DP 6, DP 21
415 Unsupported Media Type	DP 6, DP 21, DP 23
416 Unsupported URI Scheme	DP 6, DP 21
480 Temporarily Unavailable	DP 23, DP 27
485 Ambiguous	DP 6, DP 21
486 Busy Here	DP 13, DP 21, DP 25
488 Not Acceptable Here	DP 6, DP 21

Acknowledgments

Special acknowledgment is due to Hui-Lan Lu for acting as the chair of the SIN DT and ensuring that the focus of the DT did not veer too far. The authors would also like to give special thanks to Mr. Ray C. Forbes from Marconi Communications Limited for his valuable contribution on the system and network architectural aspects as co-chair in the ETSI SPAN. Thanks also to Doris Lebovits, Kamlesh Tewani, Janusz Dobrowloski, Jack Kozik, Warren Montgomery, Lev Slutsman, Henning Schulzrinne, and Jonathan Rosenberg, who all contributed to the discussions on the relationship of IN and SIP call models.

Author's Addresses

Vijay K. Gurbani
Lucent Technologies, Inc.
2000 Lucent Lane, Rm 6G-440
Naperville, Illinois 60566
USA
Phone: +1 630 224 0216
EMail: vkg@lucent.com

Frans Haerens
Alcatel Bell
Francis Welles Plein,1
Belgium
Phone: +32 3 240 9034
EMail: frans.haerens@alcatel.be

Vidhi Rastogi
Wipro Technologies
Plot No.72, Keonics Electronics City,
Hosur Main Road,
Bangalore 226 560 100
Phone: +91 80 51381869
EMail: vidhi.rastogi@wipro.com

Full Copyright Statement

Intellectual Property

The IETF takes no position regarding the validity or scope of any Intellectual Property Rights or other rights that might be claimed to pertain to the implementation or use of the technology described in this document or the extent to which any license under such rights might or might not be available; nor does it represent that it has made any independent effort to identify any such rights. Information on the ISOC's procedures with respect to rights in ISOC Documents can be found in BCP 78 and BCP 79.

Copies of IPR disclosures made to the IETF Secretariat and any assurances of licenses to be made available, or the result of an attempt made to obtain a general license or permission for the use of such proprietary rights by implementers or users of this specification can be obtained from the IETF on-line IPR repository at http://www.ietf.org/ipr.

The IETF invites any interested party to bring to its attention any copyrights, patents or patent applications, or other proprietary rights that may cover technology that may be required to implement this standard. Please address the information to the IETF at ietf-ipr@ietf.org.

Acknowledgement

Funding for the RFC Editor function is currently provided by the Internet Society.

Appendix C

XML Schema for PSTN Events

```
<xs:schema targetNamespace="urn:ietf:params:xml:ns:spirits-1.0"
        xmlns:tns="urn:ietf:params:xml:ns:spirits-1.0"
        xmlns:xs="http://www.w3.org/2001/XMLSchema"
        elementFormDefault="qualified"
        attributeFormDefault="unqualified">

  <!-- This import brings in the XML language attribute xml:lang-->
  <xs:import namespace="http://www.w3.org/XML/1998/namespace"
        schemaLocation="http://www.w3.org/2001/xml.xsd"/>
  <xs:annotation>
      <xs:documentation xml:lang="en">
          Describes SPIRITS events.
      </xs:documentation>
  </xs:annotation>

  <xs:element name="spirits-event" type="tns:SpiritsEventType"/>

  <xs:complexType name="SpiritsEventType">
      <xs:sequence>
        <xs:element name="Event" type="tns:EventType" minOccurs="1"
            maxOccurs="unbounded"/>
         <xs:any namespace="##other" processContents="lax"
            maxOccurs="unbounded"/>
      </xs:sequence>
  </xs:complexType>
```

```
<xs:complexType name="EventType">
    <xs:sequence>
        <xs:element name="CalledPartyNumber" type="xs:token"
            minOccurs="0" maxOccurs="1"/>
        <xs:element name="CallingPartyNumber" type="xs:token"
            minOccurs="0" maxOccurs="1"/>
        <xs:element name="DialledDigits" type="xs:token"
            minOccurs="0" maxOccurs="1"/>
        <xs:element name="Cell-ID" type="xs:token"
            minOccurs="0" maxOccurs="1"/>
        <xs:element name="Cause" type="tns:CauseType"
            minOccurs="0" maxOccurs="1"/>
    </xs:sequence>
    <xs:attribute name="type" type="tns:PayloadType"
        use="required"/>
    <xs:attribute name="name" type="tns:EventNameType"
        use="required"/>
    <xs:attribute name="mode" type="tns:ModeType"
        use="optional" default="N"/>
</xs:complexType>

<xs:simpleType name="PayloadType">
    <!-- The <spirits-event> will contain either a list of -->
    <!-- INDPs events or a list of userprof events -->
    <xs:restriction base="xs:string">
        <xs:enumeration value="INDPs"/>
        <xs:enumeration value="userprof"/>
    </xs:restriction>
</xs:simpleType>

<xs:simpleType name="EventNameType">
    <xs:restriction base="xs:string">
        <!-- These are the call related events (DPs). If the -->
        <!-- PayloadType is "INDPs", then the value of the -->
        <!-- "name" attribute is one of these; example -->
        <!-- <spirits-event type="INDPs" name="OCI"> -->
        <xs:enumeration value="OAA"/>
        <xs:enumeration value="OCI"/>
        <xs:enumeration value="OAI"/>
        <xs:enumeration value="OA"/>
        <xs:enumeration value="OTS"/>
        <xs:enumeration value="ONA"/>
        <xs:enumeration value="OCPB"/>
```

```
              <xs:enumeration value="ORSF"/>
              <xs:enumeration value="OMC"/>
              <xs:enumeration value="OAB"/>
              <xs:enumeration value="OD"/>
              <xs:enumeration value="TA"/>
              <xs:enumeration value="TMC"/>
              <xs:enumeration value="TAB"/>
              <xs:enumeration value="TD"/>
              <xs:enumeration value="TAA"/>
              <xs:enumeration value="TFSA"/>
              <xs:enumeration value="TB"/>
              <!-- These are the non-call related events. If the -->
              <!-- PayloadType is "user-prof", then the value of -->
              <!-- the "name" attribute is one of these; example -->
              <!-- <spirits-event type="userprof" name="LUDV"> -->
              <xs:enumeration value="LUSV"/>
              <xs:enumeration value="LUDV"/>
              <xs:enumeration value="REG"/>
              <xs:enumeration value="UNREGMS"/>
              <xs:enumeration value="UNREGNTWK"/>
         </xs:restriction>
    </xs:simpleType>

              <xs:simpleType name="ModeType">
                <!-- One of two values: "N"otification or "R"equest -->
                  <xs:restriction base="xs:string">
                      <xs:enumeration value="N"/>
                      <xs:enumeration value="R"/>
                  </xs:restriction>
                </xs:simpleType>

  <xs:simpleType name="CauseType">
      <xs:restriction base="xs:string">
          <xs:enumeration value="Busy"/>
          <xs:enumeration value="Unreachable"/>
      </xs:restriction>
  </xs:simpleType>
</xs:schema>
```

Appendix D

XML Schema
for SMS to IM

```
<xs:schema targetNamespace="http://www.iit.edu/sms-1.0"
           elementFormDefault="qualified"
           attributeFormDefault="unqualified"
           xmlns:tns="http://www.iit.edu/sms-1.0"
           xmlns:xs="http://www.w3.org/2001/XMLSchema"
           xmlns:ns1="http://www.w3.org/XML/1998/namespace">

   <!-- This import brings in the XML language attribute xml:lang-->
   <xs:import namespace="http://www.w3.org/XML/1998/namespace"
              schemaLocation="http://www.w3.org/2001/xml.xsd"/>

   <xs:annotation>
      <xs:documentation xml:lang="en">
            Describes SMS to IM Schema.
      </xs:documentation>
   </xs:annotation>

   <xs:element name="sms" type="tns:smsType"/>
      <xs:complexType name="smsType">
         <xs:sequence>
            <xs:element name="DeliveryType" type="tns:DeliveryType"
                maxOccurs="unbounded"/>
            <xs:element name="IM" type="xs:anyURI"/>
```

```
          <xs:any namespace="##other" processContents="lax"
              minOccurs="0"
              maxOccurs="unbounded"/>
      </xs:sequence>
      <xs:attribute name="Principal" type="xs:anyURI"
          use="required"/>

      </xs:complexType>
          <xs:simpleType name="DeliveryType">
              <xs:restriction base="xs:string">
                  <xs:enumeration value="Failure"/>
                  <xs:enumeration value="In-addition-to"/>
              </xs:restriction>
          </xs:simpleType>
  </xs:schema>
```

Appendix E

Raw Data for Event Manager Performance Analysis

Performance Run: Oct. 16, 2004
Database size: 1 Million entries

Host: Sun Microsystems Netra 1405 UltraSPARC-II, 4 CPU @ 440MHz; 4 Gbytes memory.

λ: Arrival rate (events/sec)
T: Total execution time (sec)
S: Average service time per event (ms)

Table E.1   Raw Data

$\lambda = 200$		$\lambda = 400$		$\lambda = 600$	
T	S	T	S	T	S
1.14	1.59	2.4	1.82	4.46	0.98
1.96	1.58	2.32	1.1	4.56	1.24
1.13	1.86	2.27	1.32	2.69	1.17
0.99	1.02	2.36	1.7	2.76	1.33
1.81	1.63	2.55	1.33	4.55	1.46
1.31	1.52	2.4	1.13	2.53	1.4
1.31	1.05	2.35	1.15	2.68	1.24
1.76	1.37	2.51	1.25	4.6	1.66
1.45	1.76	2.31	1.35	4.5	1.32
0.94	0.86	2.24	1.32	5.81	1.59
Average **1.38**	**1.42**	**2.37**	**1.35**	**3.91**	**1.34**

Appendix F

Bibliography

[3RD02]   3rd Generation Partnership Project, Multimedia Messaging Service (MMS): Stage 1, Technical Specification TS 22.140, version 5.4.0, December 2002.

[ACK99]   Ackermann, D. and Chapron, J.-E., Is the IN call model still valid for new network technologies? in *Proceedings of the 1999 International Conference on Intelligent Networks*, April 1999, n.pag.

[ARL04]   Arlein, R. and Gurbani, V.K, An extensible framework for constructing session initiation protocol (SIP) user agents, *Bell Labs Technical Journal*, 9, 87–100, 2004.

[BAC00]   Bacon, J., Moody, K., Bates, J., Hayton, R., Ma, C., McNeil, A., Seidel, O., and Spiteri, M., Generic support for distributed application, *IEEE Computer*, 33, 68–76, 2000.

[BAR93]   Barr, W.J., Boyd, T., and Inoue, Y., The TINA initiative, *IEEE Communications*, 31, 70–76, 1993.

[BER92]   Berman, R.K. and Brewster, J.H., Perspectives on the AIN architecture, *IEEE Communications*, 30, 27–32, 1992.

[BER97]   Bergeren, M., Bollinger, B., Earl, D., Grossman, D., Ho, B.-W., and Thompson, R., Wireless and wireline convergence, *Bell Labs Technical Journal*, 2, 194–206, 1997.

[BER98]   Berners-Lee, T., Fielding, R., and Masinter, L., Uniform Resource Identifiers (URI): Generic Syntax, IETF RFC 2396, August 1998, available online at <http://www.ietf.org/rfc/rfc2396.txt>

[BRA96]   Bradner, S., The Internet Standards Process: Revision 3, IETF RFC 2026, October 1996, available online at <http://www.ietf.org/rfc/rfc2026.txt>.

[BRU99a] Brusilovsky, A., Gurbani, V.K., Varney, D., and Jain, A., Need for PSTN Internet Notification (PIN) Services, IETF Internet-Draft, Work in Progress, presented at the Proceedings of the 44th Internet Engineering Task Force (IETF), March 1999, available online at <http://www.ietf.org/proceedings/99mar/I-D/draft-brusilovsky-pin-00.txt>.

[BRU99b] Brusilovsky, A., Gausmann, E., Gurbani, V.K., and Jain, A., A Proposal for Internet Call Waiting Using SIP: An Implementation Report, IETF Internet-Draft, Work in Progress, presented at the Proceedings of the 44th Internet Engineering Task Force (IETF), March 1999, available online at <http://www.ietf.org/proceedings/99mar/I-D/draft-brusilovsky-icw-00.txt>.

[BRU00] Brusilovsky, A., Buller, J., Conroy, L., Gurbani, V., and Slutsman, L., PSTN Internet notification (PIN) proposed architecture, services, and protocol, in *Proceedings of the 6th International Conference on Intelligence in Networks (ICIN)*, January 2000, n.pag.

[BUD03] Buddhikot, M., Chandramenon, G., Han, S., Lee, Y.W., Miller, S., and Salgarelli, L., Integration of 802.11 and third-generation wireless data networks, in *Proceedings of the 22nd Annual Joint Conference of the IEEE Computer and Communication Societies*, Vol. 1, March–April 2003, pp. 503–512.

[BUR04] Burmester, M. and Desmedt, Y., Is hierarchical public-key certification the next target for hackers? *Communications of the ACM*, 47, 69–74, 2004.

[CAM94] Cameron, E.J., Griffeth, N., Lin, Y., Nilson, E., Schure, W.K., and Velthuijsen, H., A Feature Interaction Benchmark for IN and Beyond, *Feature Interactions in Telecommunications Systems*, 1994, pp. 1–23.

[CAM02] Camarillo, G., Roach, A.B., Peterson, J., and Ong, L., Integrated Services Digital Network User Part (ISUP) to Session Initiation Protocol (SIP) Mapping, IETF RFC 3398, December 2002, available online at <http://www.ietf.org/rfc/rfc3398.txt>.

[CAM02a] Campbell, B. (Ed.), Rosenberg, J., Schulzrinne, H., Huitema, C., and Gurle, D., Session Initiation Protocol (SIP) Extensions for Instant Messaging, IETF RFC 3248, December 2002, available online at <http://www.ietf.org/rfc/rfc3248.txt>.

[CAR01] Carzaniga, A., Rosenblum, D., and Wolf, A., Design and evaluation of a wide-area event notification service, *ACM Transactions on Computer Systems*, 19, 332–383, 2001.

[CHA01] Chapron, J.-E. and Chatras, B., An analysis of the IN call model suitability in the context of VoIP, *Computer Networks*, 35, 521–535, 2001.

[CHI00] Chiang, T.-C., Douglas, J., Gurbani, V.K., Montgomery, W.A., Opdyke, W.F., Reddy, J., and Vemuri, K., IN services for converged (Internet) telephony, *IEEE Communications*, 38, 108–115, 2000.

[CHI02] Chiang, T.-C., Gurbani, V.K., and Reid, J.B., The need for third-party call control, *Bell Labs Technical Journal*, 7, 41–46, 2002.

[CHR01] Christensen, E., Curbera, F., Meredith G., and Weerawarana, S., Web Services Description Language (WSDL) 1.1, W3C Note, March 15, 2001, available online at <http://www.w3.org/TR/wsdl>.

[COH76] Cohen, D., Specifications for the Network Voice Protocol (NVP), Technical Report RR-75-39, University of Southern California Information Sciences Institute, March 1976.

[COH77] Cohen, D., Issues in transnet packetized voice communications, in *Proceedings of the Fifth ACM Symposium on Data Communications*, 1977, pp. 6.10–6.13.

[COL01] Colbert, R., Compton, D., Hackbarth, R., Herbsleb, J., Hoadley, L., and Willis, G., Advanced services: changing how we communicate, *Bell Labs Technical Journal*, 6, 211–288, 2001.

[COP01] Copeland, R., Presence: a re-invention or a new concept? in *Proceedings of the 7th International Conference on Intelligence in Next Generation Networks (ICIN)*, October 2001, pp. 127–132.

[COR04] Cortes, M., Ensor, J.R., and Esteban, J., On SIP performance, *Bell Labs Technical Journal*, 9, 155–172, 2004.

[CUG01] Cugola, G., Di Nitto, E., and Fuggetta, A., The JEDI event-based infrastructure and its application to the development of the OPSS WFMS, *IEEE Transactions on Software Engineering*, 27, 827–850, 2001.

[CUG02] Cugola, G. and Jacobsen, H.-A., Using publish/subscribe middleware for mobile systems, *ACM Mobile Computing and Communications Review*, 6, 25–33, 2002.

[DAS03] Das, S.K., Lee, E., Basu, K., and Sen, S.K., Performance optimization of VoIP calls over wireless links using H.323 protocol, *IEEE Transactions on Computers*, 52, 742–752, 2003.

[DIA02] Dianda, J., Gurbani V.K., and Jones, M.H., Session initiation protocol service architecture, *Bell Labs Technical Journal*, 7, 3–23, 2002.

[DIE99] Dierks, T. and Allen, C., The TLS Protocol: Version 1.0, IETF RFC 2246, January 1999, available online at <http://www.ietf.org/rfc/rfc2246.txt>.

[DON00] Donovan, S., The SIP INFO Method, IETF RFC 2976, October 2000, available online at <http://www.ietf.org/rfc/rfc2976.txt>.

[FAY96] Faynberg, I., Gabuzda, L., Kaplan, M.P., and Shah, N.J., *The Intelligent Network Standards: Their Application to Services*, McGraw-Hill, New York, 1996.

[FAY97] Faynberg, I., Gabuzda, L., Jacobson, T., and Lu, H.-L., The development of the wireless intelligent network (WIN) and its relation to the International Intelligent Network Standards, *Bell Labs Technical Journal*, 2, 57–80, 1997.

[FAY00] Faynberg, I., Gabuzda, L., and Lu, H.-L., *Converged Networks and Services: Interworking IP and PSTN*, 1st ed., John Wiley & Sons, New York, 2000.

[FCC03] U.S. Federal Communications Commission, Trends in Telephone Service, Washington, DC, August 2003, available online at <http://www.fcc.gov/wcb/iatd/trends.html>.

[FIN02] Finkelstein, M., Garrahan, J., Shrader, D., and Weber, G., The future of the intelligent network, *IEEE Communications*, 38, 100–106, 2002.

[FOS02] Foster, I., What Is a Grid? A Three Point Checklist, 2002, available online at http://www.fp-mcs.anl.gov/~foster/Articles/WhatIsTheGrid.pdf.

[FRE96a] Freed, N. and Borenstein, N., Multipurpose Internet Mail Extensions (MIME) Part One: Format of Internet Message Bodies, IETF RFC 2045, November 1996, available online at <http://www.ietf.org/rfc/rfc2045.txt>.

[FRE96b] Freed, N. and Borenstein, N., Multipurpose Internet Mail Extensions (MIME) Part Two: Media Types, IETF RFC 2046, November 1996, available online at <http://www.ietf.org/rfc/rfc2046.txt>.

[GAD03] Gaddah, A. and Kunz, T., A Survey of Middleware Paradigms for Mobile Computing, Carleton University Systems and Computing Engineering Technical Report SCE-03-16, July 2003, available online at <http://www.sce.carleton.ca/wmc/middleware/middleware.pdf>.

[GAL97] Gallagher, M.D. and Snyder, R.A., *Mobile Telecommunications Networking with IS-41*, McGraw-Hill, New York, 1997.

[GAR02] Garber, L., Will 3G really be the next big wireless technology? *IEEE Computer*, 35, 26–32, 2002.

[GAR03] Garg, S. and Kappes, M., Can I add a VoIP call? in *Proceedings of the 38th IEEE International Conference on Communications (ICC)*, May 2003, pp. 779–783.

[GBA99] Gbaguidi, C., Hubaux, J.-P., Pacific, G., and Tantawi, A.N., Integration of Internet and telecommunications: an architecture for hybrid services, *IEEE Journal on Selected Areas in Communications (JSAC)*, 17, 1563–1579, 1999.

[GEI01] Geihs, K., Middleware challenged ahead, *IEEE Computer*, 34, 24–31, 2001.

[GLI00] Glitho, R.H., Advanced services architecture for Internet telephony: a critical overview, *IEEE Network*, 14, 38–44, 2000.

[GLI03] Glitho, R.H., Khendek, F., and De Marco, A., Creating value added services in Internet telephony: an overview and a case study on a high-level service creation environment, *IEEE Transactions on Systems, Man, and Cybernetics*, 33, 446–457, 2003.

[GOO00] Goodman, D.J., The wireless Internet: promises and challenges, *IEEE Computer*, 33, 36–41, 2000.

[GUL00] Gulbrandsen, A., Vixie, P., and Esibov, L., A DNS RR for Specifying the Location of Services (DNS SRV), IETF RFC 2782, February 2000, available online at <http://www.ietf.org/rfc/rfc2782.txt>.

[GUR99] Gurbani, V.K., PSTN Internet Notification BOF (pin), paper presented at the Proceedings of the 44th Internet Engineering Task Force (IETF), March 1999, available online at <http://www.ietf.org/proceedings/99mar/slides/pin-services-99mar/sld001.htm>.

[GUR01] Gurbani, V.K., Chiang, T.-C., and Kumar, S., SIP: a routing protocol, *Bell Labs Technical Journal*, 6, 136–152, 2001.

[GUR02] Gurbani, V.K., Haerens, F., and Rastogi, V., Interworking SIP and Intelligent Network (IN) Applications, paper presented at the Proceedings of the 54th Internet Engineering Task Force (IETF), July 2002, available online at <http://www.ietf.org/proceedings/02jul/I-D/draft-gurbani-sin-02.txt>.

[GUR03a] Gurbani, V.K, Sun, X.-H., Brusilovsky, A., Faynberg, I., Lu, H.-L., and Unmehopa, M., Internet service execution for telephony events, in *Proceedings of the 8th International Conference on Intelligence in Next Generation Networks (ICIN)*, April 2003, n.pag.

[GUR03b] Gurbani, V.K. and Sun, X.-H., Services spanning heterogeneous networks, in *Proceedings of the IEEE 38th International Conference on Communications (ICC)*, May 2003, pp. 764–768.

[GUR03c] Gurbani, V.K. and Liu, K.Q., Session initiation protocol: service residency and resiliency, *Bell Labs Technical Journal*, 8, 83–94, 2003.

[GUR03d] Gurbani, V.K. and Sun, X.-H., Accessing telephony services from the Internet, in *Proceedings of the IEEE 12th International Conference on Computer Communications and Networks (ICCCN)*, October 2003, pp. 517–523.

[GUR04a] Gurbani, V.K. and Sun, X.-H., Terminating telephony services on the Internet, *IEEE/ACM Transactions on Networking*, 12, 571–581, 2004.

[GUR04b] Gurbani, V.K. and Jain, R., Contemplating some open challenges in SIP, *Bell Labs Technical Journal*, 9, 255–269, 2004.

[GUR04c] Gurbani, V.K. (Ed.), Faynberg, I., Lu, H.-L., Brusilovsky, A., Unmehopa, M., and Gato, J., The SPIRITS (Services in the PSTN Requesting Internet Services) Protocol, IETF RFC 3910, October 2004, available online at <http://www.ietf.org/rfc/rfc3910.txt>.

[GUR04d] Gurbani, V.K. and Sun, X.-H., Extensions to an Internet Signaling Protocol to Support Telecommunications Services, paper presented at the Proceedings of the 2004 IEEE Global Telecommunications Conference (GLOBECOM), November–December 2004 (accepted for publication).

[GUR05a] Gurbani, V.K. and Sun, X.-H., A systematic approach for closer integration of cellular and Internet services, *IEEE Network*, 19, 26–32, 2005.

[GUT02] Gutmann, P., PKI: it's not dead, just resting, *IEEE Computer*, 35, 41–49, 2002.

[GUT04] Gutmann, P., Simplifying public key management, *IEEE Computer*, 37, 101–103, 2004.

[HAN98] Handley, M. and Jacobson, V., SDP: Session Description Protocol, IETF RFC 2327, April 1998, available online at <http://www.ietf.org/rfc/rfc2327.txt>.

[HIL81] Hillier, F., Yu, O., Avis, D., Fossett, L., Lo, F., and Reiman, M., *Queuing Tables and Graphs*, Elsevier North-Holland, New York, 1981.

[HIL90] Hillier, F. and Lieberman, G., *Introduction to Operations Research*, 5th ed., McGraw-Hill, New York, 1990.

[HOM03] Homayounfar, K., Rate adaptive speech coding for universal multimedia access, *IEEE Signal Processing Magazine*, 20, 30–39, 2003.

[IET05] The IETF SPIRITS (Services in the Internet Requesting PSTN Services) Working Group, <http://www.ietf.org/html.charters/spirits-charter.html>.

[ITU92a] International Telecommunications Union — Telecommunications Standardization Sector, Principles of Intelligent Network Architecture, Recommendation Q.1201, Geneva, Switzerland, October 1992.

[ITU92b] International Telecommunications Union — Telecommunications Standardization Sector, Intelligent Network Distributed Functional Plane Architecture, Recommendation Q.1204, Geneva, Switzerland, March 1993.

[ITU93] International Telecommunications Union — Telecommunications Standardization Sector, Intelligent Network Physical Plane Architecture, Recommendation Q.1205, Geneva, Switzerland, March 1993.

[ITU00a] International Telecommunications Union — Telecommunications Standardization Sector, Bearer Independent Call Control Protocol, Recommendation Q.1901, Geneva, Switzerland, June 2000.

[ITU00b] International Telecommunications Union — Telecommunications Standardization Sector, The Directory: Public-Key and Attribute Certificate Frameworks, Recommendation X.509, Geneva, Switzerland, March 2000.

[ITU01]   International Telecommunications Union — Telecommunications Standardization Sector, Distributed Function Plane for Intelligent Network Capability Set 4, Recommendation Q.1244, Geneva, Switzerland, July 2001.

[ITU03]   International Telecommunications Union — Telecommunications Standardization Sector, Packet Based Multimedia Communication Systems, Recommendation H.323, Geneva, Switzerland, July 2003.

[JAI91]   Jain, R., *The Art of Computer Systems Performance Analysis: Techniques for Experimental Design, Measurement, Simulation, and Modeling,* Wiley Professional Computing, 1991.

[JAV03]   Java Community Process, JSR 116: SIP Servlet API, March 2003, available online at <http://jcp.org/ aboutJava/communityprocess/final/jsr116/index .html>.

[JIA03]   Jiang, W., Koguchi, K., and Schulzrinne, H., QoS evaluation of VoIP endpoints, in *Proceedings of the IEEE 38th International Conference on Communications (ICC)*, May 2003, pp. 1917–1921.

[KAN01]   Kanter, T., An open service architecture for adaptive personal mobile communications, *IEEE Personal Communications*, 8, 8–17, 2001.

[KAN02]   Kanter, T., HotTown, enabling context-aware and extensible mobile interactive spaces, *IEEE Wireless Communications*, 9, 18–27, 2002.

[KEM04]   Kempf, J. and Austein, R., The Rise of the Middle and the Future of End-to-End: Reflections on the Evolution of the Internet Architecture, IETF RFC 3724, March 2004, available online at <http://www.ietf.org/rfc/ rfc3724.txt>.

[KLE75]   Kleinrock, L., *Queuing Systems*, Vol. 1, *Theory*, John Wiley & Sons, New York, 1975.

[KOZ03]   Kozik, J., Unmehopa, M., and Vemuri, K., A parlay and SPIRITS-based architecture for service mediation, *Bell Labs Technical Journal*, 7, 105–122, 2003.

[LEA04]   Leavitt, N., Are web services finally ready to deliver? *IEEE Computer*, 37, 14–18, 2004.

[LEN99]   Lennox, J., Schulzrinne, H., and La Porta, T.F., Implementing Intelligent Network Services with the Session Initiation Protocol, Technical Report CUCS-002-99, Columbia University, New York, January 1999.

[LEN00]   Lennox, J. and Schulzrinne, H., Feature Interaction in Internet Telephony, paper presented at the Proceedings of the Sixth International Workshop on Feature Interactions in Telecommunications and Software Systems, May 2000.

[LEN01a]   Lennox, J., Rosenberg, J., and Schulzrinne, H., Common Gateway Interface for SIP, IETF RFC 3050, February 2001, available online at <http://www .ietf.org/rfc/rfc3050.txt>.

[LEN01b]   Lennox, J., Murakami, K., Karaul, M., and La Porta, T., Interworking Internet telephony and wireless telecommunications networks, *ACM Computer Communication Review*, 31, 25–36, 2001.

[LEN04]   Lennox, J., Wu, X., and Schulzrinne, H., CPL: A Language for User Control of Internet Telephony Services, IETF Internet-Draft, Work in Progress, October 2004.

[LEN04a] Lennox, J., Services for Internet Telephony, Ph.D. thesis, Graduate School of Arts and Sciences, Columbia University, New York, 2004.

[LIC01] Liccardi, C.A., Canal, G., Andreetto, A., and Lago, P., An architecture for IN-Internet hybrid services, *Elsevier Computer Networks Journal*, 35, 537–549, 2001.

[LOW97] Low, C., Integrating communication services, *IEEE Communications*, 35, 164–169, 1997.

[LUH00] Lu, H.-L. (Ed.), Faynberg, I., Voelker, J., Weissman, M., Zhang, W., Rhim, S., Hwang, J., Ago, S., Moeenuddin, S., Hadvani, S., Nyckelgard, S., Yoakum, J., and Robart, L., Pre-SPIRITS Implementations of PSTN-Initiated Services, IETF RFC 2995, November 2000, available online at <http://www.ietf.org/rfc/rfc2995.txt>.

[LUH01] Lu, H.-L., Subject: AOPL, electronic mail between H.-L. Lu and author, unpublished, September 2001.

[MAN99] Maniatis, P., Roussopoulous, M., Swierk, E., Lai, K., Appenzeller, G., Zhao, X., and Baker, M., The mobile people architecture, *ACM Mobile Computing and Communication Review (SIGMOBILE)*, 3, 36–42, 1999.

[MAR03] Markopolou, A.P., Tobagi, F.A., and Karam, M.J., Assessing the quality of voice communications over Internet backbones, *IEEE/ACM Transactions on Networking*, 11, 747–760, 2003.

[MEI02] Meier, R., Communications paradigms for mobile computing," *ACM Mobile Computing and Communications Review*, 6, 56–58, 2002.

[MES96] Messerschmitt, D.G., The convergence of telecommunications and computing: what are the implications today? *Proceedings of the IEEE*, 84, 1167–1186, 1996.

[MES99] Messerschmitt, D.G., The prospect of computing-communications convergence, in *Proceedings of MUNCHNER KREIS Conference*, Munich, Germany, November 1999, n.pag.

[MIL00] Milewski, A. and Smith, T., Providing presence cues to telephone users, in *Proceedings of the 2000 ACM Conference on Computer Supported Cooperative Work (CSCW)*, December 2000, pp. 89–96.

[MIL05] Miller, F., Lu, S., Gupta, P., and Arsoy, A., Carrying TCAP in SIP Messages (SIP-TCAP), IETF Internet-Draft, Work in Progress, n.d.

[MON01] Montgomery, W., Network intelligence for presence enhanced communications, in *Proceedings of the 7th International Conference on Intelligence in Next Generation Networks (ICIN)*, October 2001, pp. 133–137.

[OAS02] The OASIS Consortium, UDDI Version 2.04 API Specification, July 2002, available online at <http://uddi.org/pubs/ProgrammersAPI-V2.04-Published-20020719.pdf>.

[PAN02] Panagiotakis, S. and Alonistioti, A., Intelligent service mediation for supporting advanced location and mobility-aware service provisioning in reconfigurable mobile networks, *IEEE Wireless Communications*, 9, 28–38, 2002.

[PAP03] Papazoglou, M., Service-oriented computing: concepts, characteristics, and directions, in *Proceedings of the 4th IEEE International Conference on Web Information Systems Engineering (WISC)*, December 2003, pp. 3–12.

[PAP03a] Papazoglou, M. and Georgakapoulous, D., Service-oriented computing, *Communications of the ACM*, 46, 25–28, 2003.

[PAR05] The Parlay Group, <http://www.parlay.org/>.

[PET00] Petrack, S. and Conroy, L., The PINT Service Protocol: Extensions to SIP and SDP for IP Access to Telephone Call Services, IETF RFC 2848, June 2000, available online at <http://www.ietf.org/rfc/rfc2848.txt>.

[RES01] Rescorla, E., *SSL and TLS: Designing and Building Secure Systems*, Addison-Wesley, Reading, MA, 2001.

[ROA02] Roach, A., Session Initiation Protocol (SIP)-Specific Event Notification, IETF RFC 3265, June 2002, available online at <http://www.ietf.org/rfc/rfc3265.txt>.

[ROS97] Rosenblum, D.S. and Wolf, A.L, A design framework for Internet-scale event observation and notification, in *Proceedings of the Sixth European Software Engineering Conference*, Springer-Verlag, Berlin, 1997, pp. 344–360.

[ROS01] Rosenberg, J., Distributed Algorithms and Protocols for Scalable Internet Telephony, Ph.D. thesis, Graduate School of Arts and Sciences, Columbia University, New York, 2001.

[ROS02] Rosenberg, J., Schulzrinne, H., Camarillo, G., Johnston, A., Peterson, J., Sparks, R., Handley, M., and Schooler, E., SIP: Session Initiation Protocol, IETF RFC 3261, June 2002, available online at <http://www.ietf.org/rfc/rfc3261.txt>.

[ROS02a] Rosenberg, J., Salama, H., and Squire, M., "Telephony Routing over IP (TRIP)," IETF RFC 3219, January 2002, available online at <http://www.ietf.org/rfc/rfc3219.txt>.

[ROS04] Rosenberg, J., A Presence Event Package for the Session Initiation Protocol (SIP), IETF RFC 3856, August 2004, available online at <http://www.ietf.org/rfc/rfc3856.txt>.

[RUS02] Russell, T., *Signaling System #7*, 4th ed., McGraw-Hill, New York, 2002.

[SAT01] Satyanarayanan, M., Pervasive computing: vision and challenges, *IEEE Personal Communications*, 8, 10–17, 2001.

[SAT02] Satyanarayanan, M., A catalyst for mobile and ubiquitous computing, *IEEE Pervasive Computing*, 1, 2–5, 2002.

[SAT04] Satyanarayanan, M., The many faces of adaptation, *IEEE Pervasive Computing*, 3, 2–3, 2004.

[SCH98] Schoen, U., Hamann, J., Jugel, A., Kurzawa, H., and Schmidt, C., Convergence between public switching and the Internet, *IEEE Communications*, 36, 50–65, 1998.

[SCH02] Schilit, B., Hilbert, D., and Trevor, J., Context-aware communications, *IEEE Wireless Communications*, 9, 46–54, 2002.

[SCH03] Schulzrinne, H., Casner, S., Frederick, R., and Jacobson, V., RTP: A Protocol for Real-Time Applications, IETF RFC 3550, July 2003, available online at <http://www.ietf.org/rfc/rfc3550.txt>.

[SCH04b] Schulzrinne, H., The tel URI for Telephone Numbers, IETF Internet-Draft, Work in Progress, June 2004, available online at <http://www.ietf.org/internet-drafts/draft-ietf-iptel-rfc2806bis-09.txt>.

[SIN00]   Singh, K. and Schulzrinne, H., Interworking between SIP/SDP and H.323, Columbia University Technical Report CUCS-015-00, New York, May 2000.

[SLU94]   Slutsman, L., Lu, H.-L., Kaplan, M.P., and Faynberg, I., The Application Oriented Parsing Language (AOPL) as a way to achieve platform independent service creation environment, in *Proceedings of the Third International Conference on Intelligence in Networks (ICIN)*, October 1994, n.pag.

[SLU01]   Slutsman, L. (Ed.), Faynberg, I., Lu, H., and Weissman, M., The SPIRITS Architecture, IETF RFC 3136, June 2001, available online at <http://www .ietf.org/rfc/rfc3136.txt>.

[STA02]   Stallings, W., *Cryptography and Network Security: Principles and Practice*, 3rd ed., Prentice Hall, Englewood Cliffs, NJ, 2002.

[SUG04]   Sugano, H., Fujimoto, S., Klyne, G., Bateman, A., Carr, W., and Peterson, J., Presence Information Data Format (PIDF), IETF RFC 3863, August 2004, available online at <http://www.ietf.org/rfc/rfc3863.txt>.

[SUN05]   Sun Microsystems JAIN APIs, <http://java.sun.com/products/jain/api _specs.html>.

[TAN01]   Tang, J., Yankelovich, N., Begole, J., Van Kleek, M., Li, F., and Bhalodia, J., ConNexus to Awarenex: extending awareness to mobile users, in *Proceedings of the ACM SIGCHI Conference on Human Factors in Computing Systems*, March–April 2001, pp. 221–228.

[TEL05]   Telecommunications Information Networking Architecture Consortium, <http://www.tinac.com/>.

[THO00]   Thompson, R.A., Telephone Switching Systems, *Artech House*, June 2000.

[TSE04]   Tseng, K.-K., Lai, Y.-C., and Lin, Y.-D., Perceptual codec and integration aware playout algorithms and quality measurements for VoIP systems, *IEEE Transactions on Consumer Electronics*, 50, 297–305, 2004.

[VAN99]   Vanecek, G., Mihai, N., Vidovic, N., and Vrsalovic, D., Enabling hybrid services in emerging data networks, *IEEE Communications*, 37, 102–109, 1999.

[VEM00]   Vemuri, K., Call Model Integration Framework, IETF Internet-Draft, Work in Progress, June 2000.

[VEM02]   Vemuri, A. and Peterson, J., Session Initiation Protocol for Telephones (SIP-T): Context and Architectures, IETF RFC 3372, September 2002, available online at <http://www.ietf.org/rfc/rfc3372.txt>.

[W3C99]   World Wide Web Consortium (W3C), Namespaces in XML, Technical Recommendation REC-xml-names-19990114, January 1999, available online at <http://www.w3.org/TR/REC-xml-names/>.

[W3C04]   World Wide Web Consortium (W3C), Extensible Markup Language (XML) 1.1, Technical Recommendation REC-xml11-20040204, April 2004, available online at <http://www.w3.org/TR/2004/REC-xml11-20040204/>.

[WAG03]   Wagstrom, P., Scarlet: A Framework for Context Aware Computing, M.Sc. thesis, Department of Computer Science, Illinois Institute of Technology, Chicago, July 2003.

[WAN00]   Wang, H., Raman, B., Chuah, C.-N., Biswas, R., Gummadi, R., Hohlt, B., Xia, H., Kiciman, E., Mao, Z., Shih, J., Subraimanian, L., Zhno, B., Joseph, A., and Katz, R., ICEBERG: an Internet-core network architecture for integrated communications, *IEEE Personal Communications*, special Issue on IP-based mobile telecommunications networks, 7, 10–19, 2000.

[WEI83]   Weinstein, C.J. and Forgie, J.W., Experience with speech communications in packet networks, *IEEE Journal on Selected Areas in Communications*, SAC-1, 963–980, 1983.

[WEI91]   Weiser, M., The computer for the 21st century, *Scientific American*, 265, 91–104, 1991.

[WEI02]   Weinstein, S., Wireless LAN and Cellular Mobile: Competition and Cooperation, Technical Talk, IEEE New Jersey Coast Section, May 2002, available online at <http://www.ewh.ieee.org/r1/njcoast/events/weinstein.ppt>.

[WIL97]   Willner, R. and Lee, W.L., IN service creation elements: variations in the meaning of a SIB, in *IEEE Intelligent Networking Workshop*, Vol. 1, May 1997.

[WUX00]   Wu, X. and Schulzrinne, H., Where should services reside in Internet telephony systems? in *Proceedings of the IP Telecom Services Workshop*, September 2000, pp. 35–40.

[WUX03]   Wu, X. and Schulzrinne, H., Programmable End System Services Using SIP, paper presented at the *Proceedings of the 38th IEEE International Conference on Communications (ICC)*, May 2003.

Index

T

U